When Cars Decide to Kill

to Kill

TIME FOR SOFTWARE SAFETY LAWS

PATRICIA HERDMAN

Ethi-teque Inc.

Toronto, Canada

Patricia Herdman / Ethi-teque Inc.
One Yonge Street, Suite 1801
Toronto, Ont. CANADA M5E 1W7
http://www.glitchwatch.com
Twitter: @glitchwatch7

Book Cover Design by Ilian Georgiev at fiverr.com

When Cars Decide to Kill / Patricia Herdman. —2nd Ed.
ISBN 978-1-988470-03-0

Table of Contents

*This book is dedicated to **Betsy Benjaminson**, a whistleblower who in good conscience provided the U.S. Congress and the media with documents that helped shed light on Toyota's Sudden Unintended Acceleration (SUA) incidents as they related to car software problems.*

When asked why she put herself in hot water with Toyota (who is putting legal pressure on her), she said she realized, "I can do something meaningful in helping prevent this horrible kind of tragedy, or I can turn away, and just keep making my paycheck, and just be a little screw in a huge machine. So I made a choice that . . . I can maybe change the world with these documents. Just a little."[1]

*Automobile owners now "report that their vehicles **act as if possessed** and leave them in dangerous situations — stalled vehicles stranded without warning on the highway, fuel pumps that won't shut off, windows that open and shut, air bags that won't deploy."[2]*

—CLARENCE DITLOW, the Center for Auto Safety

Preface

During the 2014 Christmas holidays, I was at home writing a different book. It was a book for executives in the business world about software and quality management issues. I was busy writing it because I felt it was time that "the big guys" start to understand how their decisions contribute to defective, sometimes dangerous, software. Having been in the business of software quality management for over 25 years, I know first hand that because the vast majority of executives don't really "get" what it takes to build solid software, it's common for them to make decisions that compromise quality.

As part of my research, I went online and entered the search words "software glitch". Over and over again, articles kept popping up about how, for the past decade, broken software in our cars has been devastating people's lives. Clicking on article after article, I'd soon gone down a rabbit hole where I found three things that startled me:

- There are no software safety laws – car companies can build software any way they like, whether it puts the public at risk or not.

- Today's cars are controlled by software far more than most of us understand. When it all goes wrong, it's *almost* impossible to prove that it was the software, not you, which made the car go out of control.

- Although automobile companies are required by law to report known problems with their vehicles, some companies routinely – and knowingly – break this law. When they are caught flagrantly disobeying the law, there is no meaningful consequence (the maximum fine that can be set by a U.S. Regulator is less than 0.25% of a given automaker's annual profit).

How could I not have known this? I wondered.

Yes, I vaguely remembered the various stories about Toyota's sudden acceleration problems but there had been a number of conflicting news reports over the years and, because I didn't own a Toyota, I paid little attention to these confusing stories.

That Christmas I tried to put all of this out of my mind and returned to working on the other book. But as I continued researching, more and more details kept surfacing about the fact that some software in our cars is now quite deadly – and has been for years.

Incredulous, I decided to try to sort out what was going on. It took three full days of sorting through newspaper articles and scientific reports to put all the pieces together. I didn't like what I was coming to understand. Each morning I would wake up knowing that many people had died horrible deaths because of defective software, that many more would still die, and that Western democracies had not bothered to do anything about it.

Knowing what I know about how to build and test software properly – and why it's not happening in so many organizations today, whether it be within the automobile industry or elsewhere – and knowing that software is reaching into every part of our lives, I decided to put aside the other book and take the time to write this one.

I've written *When Cars Decide to Kill* for anyone who drives or just travels along in cars – just about all of us – because it is now critically important we collectively understand the negative effects software can have on our lives and do something about it. At the very least, we all have a right to understand the risks we take when stepping into a car in the 21st century.

We must immediately make every effort to protect ourselves from deadly software, especially as robot cars start driving on our roads.

This book is a call to action. There are things we can and must do because, at this moment in history, we are all equally vulnerable. Not even those who run the automobile industry are safe when a car decides to kill.

Special Note re "Software"

Please note that throughout this book I will speak to software issues but the fact is, software doesn't work without hardware and hardware doesn't work without software. Rather than keep referring to electronics, embedded electronics, computer chips, operating systems and software, I will refer to this group of technologies often simply using the term "software" for ease of writing and reading.

CHAPTER ONE

Three Bikes and a Pony

Many years ago, I was driving along a dirt road in the country with a saddle and a broken bridle on the seat next to me. I was in Mennonite country, a place where people of the Mennonite faith live without electricity or telephones or cars. They work the land, sell eggs and maple syrup and, like the Amish of Pennsylvania, travel by horse and buggy or bicycle.

At that time, I too lived in a farmhouse because our family had horses. Like the Mennonites, we happened to heat our home with wood but we also had electric baseboard heaters as backup. We lived quite comfortably with our 20th century luxuries: two telephone lines, a refrigerator, a washer and dryer, a stove, a microwave, a water heater and a few computers. Even before the widespread use of the Internet, computers were a necessity in our household because I am in the software business, consulting to some of the world's largest – and most technologically complex – companies in both Europe and North America.

That day, I was on my way to visit a Mennonite man who fixed saddles and bridles and other leather gear used on horses. That's a hard skill to find now that the "horseless carriage" has become our primary mode of transportation.

As I neared his farm, I saw a cloud of dust up ahead of me. Then I saw what was making the dust. Four young Mennonite boys, three on bicycles and one on a pony, were biking and galloping at full throttle. The boys in their suspenders and baggy Mennonite pants raced along joyfully until they all came to a full stop at the intersection.

The three boys on bikes put their feet to the ground to balance themselves and then all four boys looked left and all four boys looked right.

All clear.

The boys on the bikes sped off but the boy on the pony didn't move. His pony wouldn't budge no matter how much he tried to get it to go.

Only when the three boys on bikes had crossed the road did I understand why. That's when the pony looked left, and then looked right.

Satisfied the road was clear, the pony galloped off with his young charge.

At the time, I thought, "I wish cars were smart enough to decide when to cross the road" but this, I now understand, was a foolish wish because these days cars *are* making decisions. And, sometimes, the decisions they make kill and maim people.

Actual Software Failures

When a car ignores its driver, it does so because of software.

More and more these days, the software in your car is making decisions on your behalf and you likely don't know what they are. Worse, there are no regulations to protect you from software defects that trigger a car to decide against your will. None. Absolutely none.

People are rightly concerned about the safety of "self-driving" cars (i.e., cars driven by software) but did you know that right now – today (2015) -- your car's software and supporting electronics can cause the:

- Brakes to fail when you try to stop the car,
- Brakes to suddenly and harshly activate on their own while you're driving,
- Doors to pop open while you're driving,
- Engine to shut down suddenly while you're driving,
- Engine to suddenly speed up uncontrollably,
- Engine to speed up suddenly and, at the same time, cause the brakes not to work,
- Engine to catch on fire while you're driving or just idling,
- Car to roll away while parked and turned off because the gear popped into neutral on its own,
- Engine to turn on and off then on again and off again while parked in your garage (causing the risk of carbon monoxide poisoning in your home),
- Cruise control to turn itself on and speed up or slow down,
- Airbags to suddenly deploy when they're not supposed to,

- Airbags *not* to deploy when they're supposed to,
- Seatbelts to suddenly tighten, or to loosen when they're not supposed to,
- Windows to open and shut when you don't want them to do either,
- Steering to fail, making it impossible to control the car.

If you think I'm exaggerating, all of the above situations have been part of various automakers' recall notices that I have listed later in this book.

Today's automobile software also allows hackers to remotely:

- Turn your car off, steer the car, make the brakes work or not work, unlock the doors and honk the horn relentlessly.

Even without the hack factor, because of faulty software, drivers have found themselves hurtling along the road, trapped in a missile they can't control.

Some Software Facts

The following "natural law" of the software world underpins this book. We software testers approach our work keeping this fact in mind:

The more complex the software, the greater likelihood there is of errors in the code (known as "bugs").

The corollary to the above software law is this:

The more complex the software, the greater the need for caution when building and testing it.

And if flawed software can kill you, then we need laws to protect us. We need regulations to make software safer than it is today.

Many of you may not have had anything at all go wrong with your car's software – yet. Others, like family and friends of mine, have experienced strange software behavior that is not life threatening but is annoying and possibly a sign of things to come. For example, an acquaintance of mine who has a computer science degree drives a brand new car with "smart" braking. Driving along, his car can tell that the vehicle ahead of him has slowed down and so his new car automatically slows down, too. But when he changes lanes, his car speeds up again even if a car is stopped at a red light directly in front of him.

Fortunately, the braking software in his car still works when he presses his foot against the brakes and he has been able to use them to avoid collisions.

Understanding software, he paid close attention to this pattern and then reported it to the dealership. The dealership's response? They directed him to speak with the salesman.

Why wasn't this technology problem properly noted by the dealer and reported to company headquarters? What can a salesman do about a software problem?

In the software quality and testing business, the best way to find and fix a technical problem is to keep track of when it appears and to note the conditions and frequency under which it occurs. Usually, one hopes to find and fix software problems before the product is sold to the customer but even after it has left the test lab, responsible companies put in mechanisms to keep track of things that go wrong with their software.

Currently, the U.S. National Highway Traffic Safety Administration (NHTSA) does not mandate this type of tracking on faulty software. It should, and the public needs to be educated about what to do when they encounter a problem. As well, the regulators of other countries very much need to work with the U.S. government to ensure that the public is getting all the facts so that we know how to protect ourselves.

> *In the <u>appendices</u>, I have listed steps you can take to try to control a vehicle that has put your life at risk. Please share this information.*

Something is clearly very wrong with how our society has been dealing with complex software that can kill us. It is urgent we take the steps necessary to regulate safety-critical software properly – especially now that companies are pushing to test their "self-driving" cars on our roads.

Here in the early part of the 21st century, when crossing a road, I would trust a pony's decision-making capabilities over my car's software any day of the week.

Why?

Because at least a pony has some skin in the game . . . and because I understand how commercial software is made.

CHAPTER TWO

"Hold On and Pray"

These days, a fully loaded car is like a loaded gun.

You are driving the speed limit on the highway with your family. Suddenly, the software in your car decides to accelerate to over 100 m.p.h. and, at the same time, causes the brakes to fail. You pump the brakes, over and over again, but it doesn't matter. You and your family are going to die, and you will die because of faulty software.

This is quite possibly what happened in 2009 to Mark Saylor and his family, although it will always remain a question because the Toyota Lexus he was driving "was too badly damaged to know for sure."[3] This particular accident became widely publicized because just prior to the horrific crash, Emergency Services received a 911 call from the doomed automobile. Until that call, Toyota held firm to their longstanding claim that Sudden Unintended Acceleration (SUA) was only due to driver error, nothing more.

Drivers, according to Toyota, were mistakenly pressing the accelerator when they meant to be pressing the brakes. The company's story quickly changed, however, after the widely publicized 911 call.

When Mark Saylor, an off-duty California State Highway Patrol Officer, found that the Toyota vehicle he was driving had suddenly raced to over 120 m.p.h. and he was completely unable to control the car, his brother-in-law, Chris Lastrella (who was a passenger in the vehicle), dialed 911 and gave an account of what was happening. The final words heard from the car before all four family members died are recorded on that call. "There's no brakes," he said, "Hold on and pray."[4]

In an *NBC News* article, California State Highway officials reported that Mark Saylor "tried to make a left turn when the freeway ended . . . but he was going too fast and struck a Ford Explorer. . . The Lexus then broke through a fence and struck a dirt embankment, catapulting it through the air more than 100 feet. The vehicle landed in dense vegetation near a riverbed"[5] and burst into flames, starting a small brush fire.

Now remember: we don't know for a fact that Mark Saylor's problem was because of software issues. But we don't know for certain it wasn't.

What we do know is that evidence before the U.S. Justice Department now shows that Toyota knew about equipment problems that cause SUAs years before the Saylor family died.[6] We also know that experts have identified the software problems that cause SUAs in a Toyota vehicle where the driver was injured and the passenger died.[7] This was confirmed in the 2013 Oklahoma *Bookout v. Toyota* case.

After the Mark Saylor crash, Toyota finally acknowledged that their cars had problems with floor mats and "sticky pedals" that can cause SUA[8] but insisted their cars had no SUA-related software issues.

As you read through this book, I invite you to question that claim. Given these facts, below, ask yourself if we have unearthed the root of the problem:

- SUAs in Toyotas continue even after the recalled vehicles have had their floor mats and sticky pedals fixed, [9]
- Software defects that trigger SUAs were identified in a Toyota vehicle and this information was presented at the *Bookout v. Toyota* trial in 2013.[10]

Few people seem to know the above facts. What many people do remember about the many Toyota runaway car incidents is that in 2011 the company asked NASA to look at *some* of their software code and the space agency reported it couldn't find the software problem. What the media failed to highlight was that NASA also stated that just because they couldn't find the problem, it didn't mean a software problem didn't exist. As they pointed out:

> *"Because proof that the ETCS-I [electronic throttle software] caused the reported UAs was not found does not mean it could not occur."* [11] – NASA Report re Toyota

Plus, NASA also said:

> *"While electronic systems may reduce the likelihood of mechanical failures, they can potentially introduce*

anomalous modes not present with those mechanical systems. "[12]

. . . meaning that Toyota's electronic systems might be introducing problems that mechanical systems don't produce.

Finding and fixing software issues can be a tricky business and it will only get trickier the more our cars are laden with software. Wrapping effective safety regulations around automobile software – and actively monitoring strange behavior like my friend's "smart braking" system – is our only hope of protecting ourselves.

Getting Closer to the Facts

A common feature in newer vehicles, electronic throttle control uses sensors and computer chips to pass commands between the vehicle's gas pedal and the engine to control a car's speed instead of the simple mechanical connection used in older cars.[13] -- CNN (2011)

The facts about Toyota's software in the crash involving Jean Bookout begin to tell a different story from the claims Toyota makes about its electronic throttle control software.

Virtually all cars on the road today have electronic throttle control software but it is not something you see when you open the hood of your car. Housed in a plastic container, it is connected to your engine and controls how fast or slow your car will go based on a number of factors. If you are driving on cruise control, for example, the software in your electronic throttle control unit will take your target speed plus the incline of the road – are you going up hill or down hill? – and will adjust its messages to the engine to control the speed.

Unfortunately, in some cases, it can adjust the speed without input from either the driver or the road conditions. Sometimes, the car will ignore the driver's attempts to brake and the engine will continue to race.

Even Toyota was aware of this fact. Back in 2002, Toyota issued a Technical Service Bulletin to fix an "engine surging" problem[14] with their 2002 Camrys.

In 2007, Jean Bookout experienced a problem with her Camry as she tried to exit the highway. At the *Bookout v. Toyota* trial, the following evidence was presented:

- The Camry suddenly raced out of control,
- As Jean Bookout exited the highway, her brakes didn't work,
- At some point, she pulled the parking brake.

According to the court transcripts, evidence from the crash site confirms this:

> *"[T]here are indications that the parking brake was indeed pulled and this resulted either from the parking brake or the service brake or both in* ***a skid mark of 150 [feet]*** *leading to a crash site in a ditch [past] a stop sign at the end of the exit ramp."*[15] [Emphasis added.]

Mrs. Bookout was not pressing the accelerator when she meant to be pressing the brakes.

Despite doing all the right things as a driver, Jean Bookout was injured and her passenger, Barbara Schwarz, died.

At the Oklahoma *Bookout v. Toyota* trial, a jury also heard the testimony of embedded electronics software expert, Michael Barr.

After he and a team of six other software engineers examined Toyota's:

- Electronic throttle control software source code,
- The third party operating system it runs on, and
- The hardware it uses . . .

. . . Then tested them in a "test lab" using what is called a "simulation" test program. Finally, Mr. Barr arranged to have the software tested with two real Toyota cars, causing them to suddenly accelerate. In short, Mr. Barr was able to prove to the jury that Toyota's faulty software *had* caused the accident.[16]

The jury ruled in favor of the Bookout family because they believed the expert witnesses, their code analysis and the test results, not Toyota's claims.

> *It is significant that this was the first and only jury so far to hear any opinions about Toyota's software defects. Earlier cases either predated [Barr's] source code access, applied a non-software theory, or was settled by Toyota for an undisclosed sum.*[17] – LA Times

In short, the jury believed that "the software connected to a midsize Camry's electronic throttle-control system were the cause of the unintended acceleration."[18]

A fair chunk of the overall Toyota story around the SUA situation deals with floor mats and sticky pedals, but lurking somewhere under the hood is also faulty software, as embedded electronics experts have now proven. As we go further into the Toyota saga and the misleading and often completely inaccurate messages we got from the company all along, remember two things:

- It is easier to uncover physical design problems like floor mats and sticky pedals than it is to uncover software problems,
- It took U.S. regulators ten years to untangle what was going on at Toyota with the floor mats and sticky pedal problems.

So without proper reporting mechanisms in place regarding software quality, even if we manage to survive a crash where the car ignores our commands, how can we defend ourselves if we are deemed guilty of reckless driving until we can prove that it was flawed software?

Even car companies, themselves, are sometimes challenged to find the actual source of a software problem, as we can see by how Toyota handled the speed control issues experienced by the Japanese Imperial Family's luxury car.

Toyota Fixes Imperial Family's Car Software

Although Toyota does not acknowledge SUA software errors might exist for any of its other customers, in 2008 it took software issues very seriously for one special customer: the Japanese Imperial family.

The Japan Times reported:

> *[When] the Japanese Imperial family experienced "speed control" problems with their luxury Toyota in 2008, the engineers looked into the problem, thought it might very well be a "ghost in the computer" and out of concern for the imperial heir to the throne, Toyota "**replaced the gas pedal,**

the throttle system and the engine computer at its own expense." [19] [Emphasis added.]

It is very possible that hundreds of people around the world have perished as a result of Toyota's defective software, but because it happens in small doses – a single fatality here, an entire family there – and because the default explanation is always driver error, it is easy to lose sight of what's really happening.

Toyota took the necessary steps to protect the Japanese Imperial Family's heir to the throne, but everyone's child deserves protection from faulty software, not just royalty.

The U.S. Government Takes Notice

We would likely have not known much about Toyota's special software update to the Japanese Imperial Family's car had it not been for the courage of Toyota whistleblower, Betsy Benjaminson. A Japanese-to-English translator hired by a New York law firm representing Toyota on a case, she had access to hundreds of documents and read the memos and emails going back and forth between the Japanese engineers and management. She also read what the PR department was telling management.

There were differences, she felt, between what the engineers were saying and what the PR department was communicating to the public. "Benjaminson felt that the documents in her hands proved that Toyota engineers knew there was a serious problem with their cars causing sudden acceleration, something well beyond oversized floor mats." [20]

As reported in *The Times of Israel:*

"She tried to dismiss it at first, passing Toyota's response off as standard corporate spin, a major company trying to put its best foot forward. 'But when I heard the top PR guy say things like, 'We will crush our opponents in the media,' I thought, 'This is over the line,' she said."[21]

Reporting that Toyota finally admitted it had known about the floor mat and sticky pedal problems long before the very public Mark Saylor crash in 2009, *The Washington Post* highlighted that in 2014 the company also admitted it had:

"...lied to regulators, Congress and the public for years about the sudden acceleration of its vehicles . . . [U.S. prosecutors] say Toyota's efforts to conceal the problem and protect its corporate image led to a series of fatalities that could have been prevented." [22]

Toyota was ultimately criminally charged and fined by the U.S. Justice Department $1.2 billion in March 2014 – 6.7% of their profits for that year.[23] "The facts of the case describe a level of coordinated lying and greed that warrants stiffer punishment," say critics, and the "fine means little as long as no executives face jail time."[24]

No executives face jail time.[25]

NASA and the Software Code: SUA Update

Turning back to the software issues that lurk in our cars today, automakers are beginning to warn that software problems in electronic throttle control units can indeed cause SUAs (Honda being the first to acknowledge this[26]).

But why could The Barr Group find the software problems when NASA couldn't find them in the Toyota code?

Michael Barr, of The Barr Group, addressed the possible reasons why NASA was unable to find what his team of software engineers had been able to pinpoint. As he reported after the *Bookout v. Toyota* case in October, 2013:

> *In our analysis of Toyota's source code, we built upon the prior analysis by NASA. First, we looked more closely at more lines of the source code for more vehicles for more man months. And we also did a lot of things that NASA didn't have time to do, including reviewing Toyota's operating system's internals, reviewing the source code for Toyota's "monitor CPU", performing an independent worst-case stack depth analysis, running portions of the main CPU software including the RTOS in a processor simulator, and demonstrating – in 2005 and 2008 Toyota Camry vehicles – a link between loss of throttle control and the numerous defects we found in the software.*[27] [28]

Basically, they looked at a lot more code than NASA, ran simulated tests that NASA didn't do and even arranged for testing the code in two different Toyota Camry vehicles to recreate an SUA, which NASA also didn't do.

Plus, The Barr Group had more actual software engineers looking through the code than NASA did and more time to conduct their investigation.

Putting the Pieces Together to See
the Whole Picture

When software is suspected as a possible cause for a crash, the subsequent investigations are very complex (as you can see from the fairly straight-forward excerpt from Michael Barr's blog). It's no longer a simple case of, "Oh, look, there's a broken cable!" Highly skilled experts are now required to pick apart very complex software code and try to make sense of it all. Conflicting newspaper stories drag on for years and the facts get blurred behind both the confusing storyline and the pull of the automotive industry's media power. It becomes very difficult for the average person to figure out what's really going on, especially when it is so easy for an automaker to claim "driver error".

> *"Knowing what to look for and when to pursue electronics as a candidate cause of unsafe vehicle behaviors will be increasingly important to NHTSA."*[29] – National Research Council report for the U.S. Transportation Research Board, 2012

While it is a good thing that America's National Research Council thinks that the U.S. agency in charge of regulating the car industry should be paying more attention to technology problems, the fact remains that if drivers and citizens don't get involved with how car software is regulated, only the automakers will have a say about it.

Plus, for those of us who live outside the U.S., we need to reach out to our own country's regulators.

As citizens, we need to ensure that carmakers produce the safest products possible and we need to create a transparent safety certification process. We must stop making it easy for companies to blame the driver for their own defective software and insist they build mechanisms that accurately support or refute driver claims of faulty software.

Now that companies like Google and even Toyota are scrambling to put self-driving cars on our roads (meaning a car that makes all driving decisions by software, not by you), it's time to make sure that automotive software is built – and tested – right.

CHAPTER FOUR

The Power of "AND"

Safety critical software should never be built the way it's built today by simply using optional coding standards. Safety critical software should be built according to a set of safety regulations that all companies must follow.

We are blindly racing down a technology road that has not been designed for our individual and collective safety. At this moment in history it is our duty to change that.

I can already hear those in the media who are automaker-friendly and the misleading company spokespeople start twisting the key message in this book. "But software has saved lives," they'll say. "It's not all bad."

Or worse, "Yes, bad things happen but technology's gotten ahead of us and there's nothing we can do."

Why do I say these statements will be made? Because I've already come across them when reading about people who have tried to get the automakers to do the right thing.

For the record, I am not saying that all automobile software is bad, but we can do a lot better at building and monitoring the software we have right now. As for "technology getting out ahead of us", it has always been way out ahead of regulators but that doesn't mean we should give up trying to make technology work for us rather than kill us. If we sit back and let technology set the course of our lives, we are saying that it is all right for the handful of people who control that technology to control our democracy.

Technology should never trump democracy. Society needs to prepare for the technology looming on the horizon. It's our job as citizens to work with our own governments to define how we wish to use technology and what risks we're willing to take, and then make technology accommodate that framework.

So, when looking at the software that is running our cars today, this book is to help people understand that:

- Yes, automobile software can be a good thing *AND* sometimes it can be deadly *AND* we have the power and the technical know-how to make software safer.
- We just need the political will *AND* we have had the political will to protect people when other new technologies burst on the scene, unregulated. It is time to do that now.

Another "*AND*" to point out is that automobile companies have historically been very good at building and testing software in the 20th century *AND* yet when their executive team focuses more on profits and market share, quality tumbles and people suffer.[30] [31] This is true of the auto industry in terms of parts and engineering processes as well as true of commercial software development processes.

I am not opposed to software in automobiles. In many cases, key software features can contribute to increased safety. Recently, while I was driving in the far left lane on a snowy day, someone in the middle lane decided he wanted to hang a sharp left immediately in front of me, so he did. My excellent Anti-lock Braking System (ABS) helped me miss T-boning this foolish driver by about one inch. Had I not had ABS, there would have been an ugly crash. So, yes, I know the value of some types of automobile software.

Plus, I have gladly been in the software industry for over 25 years and have watched it *begin* to "grow up" – but it still has a long way to go. Right now, the public has absolutely no protection if companies choose to foist deadly software on their customers – until something very bad happens, and then it can take many years and mountains of heartache to uncover the true story. And during those years, people continue to die.

It shouldn't be that way.

Our cars are more like robots

There is no such thing as a "driverless" car. *Something* must drive a car. A hunk of metal can never just drive "all by itself". If a human is not driving a car then tens of millions of lines of software code are driving it.

More and more our cars are now functioning as robots – not robots with cute names and arms and legs but machines "capable of carrying out a complex series of actions automatically, especially one programmable by a computer."[32] This is the actual definition of a robot.

As Ryan Calo, an assistant law professor at the University of Washington who writes extensively on the subject of robotics and the law, pointed out in a recent paper:

> *Robotics is shaping up to be the next transformative technology of our time. And robotics has a different set of essential qualities than the Internet. Robotics combines, arguably for the first time, the promiscuity of information with **the capacity to do physical harm.** Robots display increasingly emergent behavior, permitting the technology to accomplish both useful and **unfortunate tasks in unexpected ways.**[33]* [Emphasis added.]

In the media these days, you will hear people in the automobile software business wax on about how much safer the "driverless" car of the future will be because it will make fewer mistakes. I have a few questions:

- Where is the scientific evidence to confirm that the complexities a driver (or even just a pony) manages on the roads today can be completely absorbed and interpreted correctly by a computer?
- Is it *ever* acceptable to ignore the additional rigors required to build safer software – and who gets to decide that?

Automakers and software companies will point to the improved safety of computer-controlled subways in Paris, France as an example of the benefits of robot cars. The fact is, subway trains do not cross busy intersections or need to deal with bicycles rolling along next to them on the tracks. They just need to make sure they don't bump into the train in front of them and that they stop at the correct, pre-determined subway station.

There is another extremely important question that I have not heard anyone speak about and yet it is critical that people – not corporate powers – answer this question democratically. Let us say that we have arrived at the point where the software in our cars has proven to be safer than your average bad driver (and I believe this day will come, but it will come much later in the 21st century than corporate publicity campaigns promote). *All software comes with risk* and this risk increases when:

- Complexity is introduced and
- Companies do not have to follow any pre-determined safety rules (i.e., when standards are "optional" or just a set of "guidelines")

This means that, *at times, excellent drivers will occasionally suffer the consequences of software errors while bad drivers will benefit from software that intercepts their poor judgment.*

So the question is:

- Does our society support putting good drivers at risk while protecting bad drivers?

If so, what level of risk do we support?

And if we do support putting good drivers at risk, how do we ensure that we reduce the risk of faulty software hurting good drivers?

Companies like Google and Toyota do not have the right to answer these questions on their own. We need a regulatory framework to answer these and other questions related to the use of robots in the real world.

In short: I support the use of software in automobiles *AND* I support regulating the development and certification of safety critical car software, setting design and security standards that apply equally across all vehicles on our roads – especially now that cars *today* are already behaving in many ways like the robot cars of tomorrow – except today we still have steering wheels and brakes so that good drivers can try to intercept bad software decisions.

In the next part of this book, I would like to help you understand how software makes decisions as well as understand the kinds of software problems car companies now warn us about through their recall notices.

CHAPTER FIVE

How Your Car Makes Decisions

Some cars have up to 70 separate electronic systems, running on 20 million lines of code. The computers are in control.[34] – The National Post (2012)

We do have valuable automobile software such as Anti-lock Braking Systems (ABS) that can keep cars from fishtailing and swerving out of control when the wheels encounter slippery driving conditions. But what does ABS matter if a software or electronic problem causes all braking and tracking control functions to fail?

And it doesn't have to be that way. Systems can be designed so that there are special features that override electronic failures and more effectively block hackers from gaining control of our cars.

As a software quality expert whose career has been spent helping very large, technologically complex organizations do the right thing, I have been alarmed to learn that the public is more willing to forgive technical errors that cause bad things to happen than they are willing to forgive human error. It makes my work trying to convince executives to do the right thing just that much harder.

When an entirely avoidable software error wreaks havoc on a company's customer base, the executives at that company know it will be quickly forgiven because they can claim it was a technical "glitch".

But why, I wondered, do people forgive technology so easily?

It seems the average person does not understand that a technology glitch is not the result of "spontaneous generation" but the result of simple programming mistakes combined with a series of human decisions and actions – all of which can be changed in order to reduce the number of technology problems that a company "delivers" to its customers.

All technical errors started out as human errors.

In this chapter, I'm hoping to help you understand a little bit about the fact that software is a "real" thing, built by humans.

Software and Embedded Electronics

Until I started writing this book, I didn't realize there are between 30 and 100 "embedded electronics" that tell our cars what to do and when to ignore us.[35]

These devices are tucked into the dashboard, behind the engine and in the ceiling and doors of our cars. Each one controls something our car does when we press a button or encounter an accident. It can also malfunction and do something it isn't supposed to do.

What is an embedded electronic device? Here's an image of one and you'll immediately recognize it as something similar to what sits inside your computer. These devices are run  by software, and when you have dozens of these things inside your car, it can get pretty complicated.

Until emissions controls laws came into effect, cars were based on old-fashioned mechanics. They were machines and you as the driver had to do all the thinking for it. But that changed in the 1980s.

At an electronics engineering "Vehicular Technology" conference in Michigan in 1980, GM presented a paper on its electronic control system (ECS) being designed to "include integral self-test and diagnostic aids" including "fail-operational redundancy and control software, self-test hardware, diagnostic fault isolation and self-test software." [36] For 1980, these features were very much ahead of their time.

The problem these days is that designs and software concepts from the 1980s and 1990s did not foresee the challenges of the 21st century.

How Software Basically Works

At the end of this chapter, I've provided some sample software code so that you can see what software "looks" like.

Fundamentally, all software is designed to make decisions based on certain conditions. It's the "if, then, else" principle, and how it works is that you write code to say "if this is happening, then do this, else do this other thing."

For example, say I decide to build a robot that picks fruit out of a bowl each day and I want it to pick only all the yellow fruit, and only on Tuesdays and Thursdays:

- IF day begins with "T"
- THEN pick all yellow fruit from bowl
- ELSE pick all other fruit from bowl that is not yellow

In this case, I've told the robot to pick yellow fruit on Tuesdays and Thursdays and all of the other fruit if it's not a Tuesday or Thursday.

That's pretty straightforward. But what if you wanted something a bit more complicated? What if you also wanted the apples and grapes picked, but only on Wednesdays, and all the other types of fruit picked on the other days?

- IF day begins with "T"
- THEN pick all yellow fruit from bowl
- ELSE
  - If day begins with "W"
  - THEN pick apples and grapes from bowl
  - ELSE pick all fruit that is not picked on days beginning with a "T" or a "W" from bowl

This is how software makes decisions, only software is written in special coding languages to make it easier to write software fast and so that the computer's Central Processing Unit (CPU) can process the directions more efficiently.

In the above case, if I wanted the robot to pick bananas only on the first Tuesday of the month, the code gets a lot trickier. That's how it is with automobile software, too. Programmers have to account for a lot of situations and then code in (tell) the software what decisions to make under certain conditions.

Sometimes, the programmer is unaware that there is a condition to consider, like interference from radio waves being emitted from other cars or devices. In these cases, radio waves can trigger the wrong software code to "kick in". If we go back to the fruit-picking program, it's like saying that a radio wave could mistakenly trigger the robot to "think" it was a Tuesday when in fact it was Friday. In this case, the robot will do the wrong thing.

Just imagine how many "if-then-else" type statements must go into running a car.

Later in the book, I will explain how to build software badly and how to build it better. For right now, I'd like you to "see" some software.

"Visible" Software – Take a Look

Here is a sample of embedded electronics software that I borrowed from a programmer. [37] It is written in a software language called "C".

As you scroll through these lines of code – code that performs a simple task – ask yourself this: How easy would it be to make a mistake?

The answer? Pretty easy. It happens all the time.

"Bugs" are found and fixed during the entire process of developing software. It's normal, and to be expected. But if companies don't follow the right procedures or if they rush their programmers, some (or lots) of bugs don't get caught during testing. That's when software bugs move from the developers' computers to your car, and that's when the situation becomes serious.

Even though, as you drive along, you cannot see your software code, it is a real thing. Someone builds it and puts it in cars. As you read about how software bugs sometimes take control of our cars, making them seem like they're "possessed", it helps to understand what is actually making the cars drive a little (or a lot) crazy.

To be fair, there is more to the software failures than just bad code. There's bad design and faulty hardware . . . all of which are supposed to be tested a certain way, separately and together. However, as I've mentioned, I will use word "software" to encompass the other technical "stuff" that goes with it.

When glancing through the sample software program, remember that our cars can contain *millions* of lines of code. That's a lot of links in a chain. That's a lot of places where something can go wrong.

The following example shows a simple bit of program code that can make decisions in your automobile. This code is a bit easier to read because the developer included comments. Comments are explanations about the code written in human, not computer, language. When a developer tucks comments into software code, it is easier for other software developers to understand what the software is supposed to do and it also helps humans avoid coding mistakes.

Unfortunately, there are lots of programs that don't include this basic quality management practice.

And, naturally, Japanese cars have software code comments written in Japanese.

As you can see, even without Japanese characters, just a few lines of well-written code can be very complex.

SAMPLE VEHICLE CODE:
C PROGRAMMING LANGUAGE

```
QString Automon::getVin()

{

if (m_serialHelper->isRunning())

{

#ifdef DEBUGAUTOMON

qDebug() << "Attempted to get VIN but SerialHelper is currently monitoring. Returning failure message";

#endif

return QString("Result could not be obtained. You must stop monitoring first!");

}

else

{

Command vinCommand;

vinCommand.setCommand("0902");
```

```
m_serialHelper->sendCommand(vinCommand);

QString returnedBuffer = vinCommand.getBuffer();

/* Remove spaces to make it easier to parse */

QRegExp rx( " " );

returnedBuffer.replace(rx, "");

/* Split each Line */

QStringList lineList = returnedBuffer.split("\x0D");

/*Next for each line, we will cut out the "49020n" so simply
asub string from 6 to line.size()*/

QString fullLine;

for (int i = 0; i < lineList.size(); i++)

{

QString thisLine = lineList.at(i);

thisLine = thisLine.section("",7,thisLine.size());

QRegExp nullFinder("00+");

thisLine.replace(nullFinder,"");

fullLine += thisLine;

}

if (fullLine.size() % 2 !=0)

{

#ifdef DEBUGAUTOMON
```

34

```
qDebug("The returned string for VIN was not of equal
bytes. Error");

#endif

return QString("The Returned Number of Bytes for VIN was
not even. Read Error");

}

QString vinNumber;

for (int i=0; i < fullLine.size(); i += 2)

{

QString hexByte = fullLine.section("",i+1,i+2);

vinNumber += QByteArray::fromHex(hexByte.toAscii());

}

return vinNumber;

}

}
```

Don't understand that?

Don't worry. The above example is "Greek" to me, too, and I'm in the software business …. But I'm in the software business as someone hired to *break* the software and find bugs, not as a programmer. I do know that if the programmer left out a little bit of that code (like, for example, forgot one bracket '}'), the code would now have a "bug" in it and might not work at all . . . or it might do something *wrong*, do something the programmer had not expected.

But it's not important that you or I understand the above software example. What's important is that you can see that *software is real. It is there in your car.*

And every time you drive your car, software is telling it to do something.

All of us, as citizens and as people who drive or are passengers in automobiles, have a right to know that the software is telling the car to do the right thing at the right time.

In short, we have a right to know that our cars are making the correct "if-then-else" decisions.

Some Software Testing Facts

Before stepping through the rest of the book, it is helpful to remember these three things that software testers know when they're looking for bugs:

1. (Again) -- The more complex the software, the more likelihood of defects (i.e., software "bugs" or "glitches") and the more time you need to find, fix and re-test the software.

2. Wherever you find the most defects, keep focusing your attention there because there are going to be more (where's there's smoke, there's fire).

3. When building safety critical software systems, a much more rigorous – *and much slower* – quality management process must be applied.

These principles apply no matter what type of software you're testing. In my opinion, Toyota's cars demonstrate this. As highlighted in the MIT Sloan Management Review report, "What Really Happened to Toyota?":

Although Toyota's Lexus and Prius models accounted for less than 25% of its sales in 2010, they were among the most technologically complex products and were involved in more than half of the number of recalls. [38] – MIT Sloan Report

Quite simply, complexity increases the chance of failure.

As we move deeper into the 21st century with ever increasing levels of software complexity in our cars, we need to be vigilant about this fact.

CHAPTER SIX

Actual Car Software Failures
(Variations on a Theme)

For the first time [2014], an automobile company [Honda] has conceded that a software glitch in electronic control units could cause cars to accelerate suddenly, forcing drivers to scramble to take emergency measures to prevent an accident.[39] – Junko Yoshida, Electronic Engineering Times

Sudden Unintended Acceleration is a problem for more than just Toyota vehicles. Honda and other car companies are discovering that customers have been experiencing SUAs with their cars, too.

Beyond the Sudden Unintended Acceleration situation, automakers are now issuing recalls due to a number of software and electronics failures in their vehicles.

There are three basic types of recalls:

1) Mechanical failures (parts, poor workmanship)

2) Software and electronic failures that trigger mechanical problems and

3) Mechanical failures that cause software and electronic failures to trigger mechanical problems.

At present, recall notices are not clearly subdivided into the three categories listed above, but I believe they should be. Understanding the source of various types of failures helps automakers zoom in on common problems and fix them. As well, it informs the media and the public about common risks in the vehicles we drive, and the source of those problems. Patterns of defects in software hint at other types of problems. As well, at times a software fix can create a new problem and it is critical that this type of information be monitored.

Because the focus of this book is on 21st century automobile safety issues – things that affect the safe functioning of vehicular software and electronics, not strictly mechanical failures – I am listing just a sample of recalls relating to items 2) and 3), above. In the first chapter, I summarized a series of actual failures caused by defective software that are now part of various automakers' recent recalls. You can jump back to read them again by clicking here.

2014 was a record year for recalls. Automakers issued 64 million recall notices, though not all were due to SUA or to software. Some were just plain old-fashioned mechanical problems. The previous record for the highest number of recall notices was set ten years earlier, in 2004: 30.8 million.[40]

When reading through the complaints registered with the NHTSA that often lead to the recalls, those reporting that cars were "behaving" strangely often have a different tone than the ones reporting simple mechanical failure. The words "I'm afraid" routinely appear in the complaints that are either clearly or likely related to software failure. Excerpts, below, are from complaints to the NHTSA about engines shutting off while driving or the car speeding up and the brakes not working.

- "**I'm afraid** of driving this car now, because the engine behaves erratically." NHTSA #10676407
- "**I'm afraid** to drive. If/when this happens again I hope I'm not on the highway or any road actually as the chance of an accident is very high. I don't feel safe. I have never had something like this happen to me, so I honestly don't know how to proceed." NHTSA #10672549
- "**I'm afraid** the car is going to stall [again, for the 5th time] while in motion and get rear ended." NHTSA #10615519
- "**I'm afraid** to drive this car!" NHTSA #10511114

One Mercedes owner complained that the car would stop and become inoperative on the freeway and local roads. When it happened three times in one week, the vehicle was taken to the dealer who "fixed" it with a CPU software update but it kept happening. In fact, the same day it received a program update, the vehicle stopped on the freeway and, as noted in the complaint, the driver "was almost hit by a UPS commercial truck and [would be] dead on that day."

The complainant went on to say that the car was towed to the dealer and because they didn't know what the problem was, they were just going to "change the whole CPU this time."

"Obviously," reported the complainant, "it is a very serious safety problem! *I am afraid to drive the car again. I am so young I don't want to die only because I drive a Mercedes-Benz.*"[41] [Italics added.]

In this chapter, I have summarized *only some* of the ongoing software-related recalls that I also list on GlitchWatch.com for reference. One of the recalls in this list is actually related to a security flaw in BMWs. In _The Hack Factor_ chapter, I write more fully about weak design features that support the hack-ability of our vehicles.

What is of particular concern to me is that it seems that software-related recalls are increasing in frequency. In the software testing world, that's really, really bad news as it indicates a potential "house of cards" set of situation – one defect triggering a host of dormant software bugs.

Software- and Electronics-Related Recalls

The following series of software glitches and electronics failures have actually caused or could cause serious problems for people and have been reported in the media. They are listed in alphabetical order by automaker and do not include all of the software- and electronics-related recalls nor all of the car companies that have software problems.

BMW (2015) issued an "over the air" fix for a "security flaw that could have allowed hackers to unlock the doors of up to 2.2 million Rolls-Royce, Mini and BMW vehicles."[42] Note that it was not BMW that found the flaw, but the German Motoring Association, ADAC. The association's technology president, Thomas Burkhardt, said they "waited for BMW to drop a patch before revealing the flaw."[43]

BMW (2012) issued a recall for 7,485 BMW Series 7 sedans "over a software glitch that can cause the sedans' doors to pop open unintentionally."[44] Driving on a bumpy road or even when an occupant touches the door could cause the door to pop open.

Ford (2014) – issued a recall for 1.1 million SUVs due a problem with power steering because at times "the computer cannot tell the driver's steering movements" and as a result disables the power steering function.[45] Ford said that it knew about the steering problems since 2009 but did not issue recalls even though there are at least 20 confirmed crashes and 8 injuries relating to the sudden loss of power steering while driving.[46]

In 2015, the NHTSA opened an investigation into Ford's handling of the problem, noting that "*the software update itself may in fact cause further issues with the affected vehicle's power steering, causing it to fail,* and ultimately requiring replacement of the torque sensor or entire steering column."[47] [Emphasis added.]

GM (2014) issued a recall regarding Chevy "Tahoes that can *randomly shift into neutral and roll away* from their parking spot or lose power on the highway . . . [L]eaving the Tahoe parked didn't solve the problem. The electronic glitch in its four-wheel-drive system can shift it into neutral even when the engine wasn't running . . . the glitch can also shift the trucks into neutral when they're running, leaving you coasting down the road . . . [I]magine it happening while you're in the middle of several lanes of heavy traffic. Or climbing a long incline, like countless hills, mountains and bridges across the U.S."[48] [Italics added.]

- Of special note in the above recall is that this "problem arose during the final stages of development. *GM tried to figure out the cause, couldn't decide whether it was a random anomaly or a systemic problem and moved on.*"[49] [Italics added.] In other words, GM couldn't figure out why the truck would suddenly roll away while parked, or why the engine would stop when driving, but they decided to start selling the truck anyway.

- Mark Phelan, the author reviewing the vehicle in *The Detroit Free Press* suggested to GM, "If you don't know whether a vehicle rolled away because of a rogue glitch or a problem, that's a pretty fair definition of a problem. Do not pass GO, do not begin selling the vehicles."[50]

GM (January 2014) recalled 370,000 trucks sold in North America due to a software glitch that *causes engine to overheat* while idling [and while driving, it turns out]. GM has confirmed 8 fires. [51] "A software malfunction makes an idling truck heat up to levels that can *start a fast-moving fire*, melting plastic parts and leading to total destruction of the vehicle, according to GM."

Houston wrestling coach Allen Paul was actually driving when this happened. Ironically, thirty minutes after receiving the recall notice about engine fires in idling trucks, his 2014 Chevy Silverado burst into flames as he drove along the road. He was able to leap from his vehicle and was unhurt, but his car was totaled.[52]

GM (September 2014) recalled 221,000 after the NHTSA investigated "allegations of inappropriate autonomous braking while driving." In other words, the *cars were braking all by themselves* while being driven. Dealers will fix the situation with a software update.[53]

GM (2014) recalled 50,571 2013 Cadillacs due to a "three to four second lag in acceleration due to the transmission control module (TCM) programming." The recall noted that this may increase the risk of a crash.[54]

GM (2014) recalled 19,225 vehicles to replace the software module that controls windshield wipers. If the battery dies and is then jump-started, the wipers will be inoperative.[55]

GM (2014) recalled 16,939 Cadillacs because "vibrations from the drive shaft may cause the vehicle's rollover sensor to command the roof rail bags to deploy . . . [increasing the] risk of crash and injury to the occupants."[56]

GM (2014) delayed deliveries of 2015 Chevy Colorados and GMC Canyons in order to fix an air bag flaw two weeks after beginning shipment. Although GM indicated the source of the problem was a wiring error that will cause the driver-side airbag not to deploy correctly, it will issue a software fix to correct the problem.[57]

GM (2014) recalled "140,067 Chevrolet Malibu sedans from the 2014 model year for a problem with the electronic brake control module that can increase the risk of crashes. GM said dealers will reprogram the control module." GM indicated that it is aware of 4 crashes that may be related to this software problem.[58]

GM (2011) recalled 50,500 Cadillacs due to an airbag-related software problem that puts the passenger-side back seat passenger at risk.[59]

Honda (2008) "changed the crash parameter for door-closing force in the electronic control unit's software code to reduce the incidents of inadvertent side air bag deployments." In 2014, the NHTSA was looking into whether a recall would be required. The regulator had received a number of complaints regarding air bags deploying while parked. In one case, when a woman closed the car door, "the passenger-side curtain air bag deployed, striking [her] 9-

year-old son . . . The boy suffered a concussion and had blood behind the ear, according to the complaint."[60]

Honda (November 2013) recalled "344,000 Odyssey minivans in the United States to fix a software defect that may lead the vehicle to suddenly and harshly brake without the driver pressing the brake pedal." Although the recall was made in 2013, the company said it could not fix it until some time in 2014.[61]

Honda (October 2014) recalled 175,356 vehicles due to a problem in the engine control unit that has caused vehicles to move or speed up abruptly (sudden unintended acceleration). *The company acknowledged that it was faulty software causing the problem.*[62] [Italics added.]

Hyundai (February 2015) recalled 200,000 vehicles "due to a flaw in the vehicles' electronic power steering system, the technology that makes moving the steering wheel easier. In some cases, a control unit component senses an error in the system and disables that assistance. . . Hyundai initially determined that the problem, which first became apparent in 2010, would not require a safety recall." Five years later, in 2015, NHTSA issued the safety recall.[63]

Jaguar (October 2011) recalled 17,500 cars when one of their employees discovered a software problem with its engine control unit. In this case, cruise control could only be turned off if the engine is turned off.[64]

Mazda (April 2014) recalled 33,000 Tribute SUVs due to a "problem on the power steering control module which will either need its software updated, or dealers might have to completely replace the power steering control module."[65] (Any Tribute recalls would also affect Ford Escape).

Mazda (April 2014) recalled 5,700 vehicles because under certain conditions, according to the NHTSA, the Power Control Module (PCM) incorrectly assumes failure of the charging system. When that happens, "the vehicle will stop charging and could result in poor acceleration, loss of steering assist and windshield wiper operation, and a possible engine stall, increasing the risk of a crash."[66]

Nissan (2011) issued a service bulletin to correct a problem with 5,300 of its battery-operated Leaf vehicles due to a software problem that would "keep some of them from restarting after an air conditioning sensor was activated and the vehicle turned off."[67]

Nissan (November 2014) recalled over 14,000 hybrid cars built in 2014 "because the electric motor may stop running due to a communications error between the motor inverter and transmission control module."[68] (Remember, when the motor stops, the airbags can't work.)

Tesla (2014) was told by the NHTSA to recall vehicles due to battery fires while charging. "Tesla issued an over-the-air software update that enables a charging Model S to reduce amperage if heating is detected . . . Bloomberg News reported that at least six incidents of plugs smoking or melting had been discussed on a Tesla owner website."[69] In one case, a house fire that had started in the garage was attributed by the firefighters to either faulty wiring in the house or to the Tesla vehicle that was having its battery charged in the garage.

Toyota, Honda and Fiat Chrysler (January 2015) have recalled 2.12 million older vehicles because of electronic defects that can cause airbags to inadvertently deploy, causing physical injury including eye damage. The airbag electronic "module is subject to electrical interference from other components in the car that can cause it to trigger one or more air bags [to] spontaneously [deploy while driving even if there has not been a crash]." Although earlier recalls in 2012 and 2014 had tried to fix the electronic problem, some cars that been repaired under those recalls "subsequently experienced inadvertent air bag deployments, prompting this latest campaign." [70]   The recall "also addressed seat-belt tighteners that could incorrectly activate."[71] [72]

Toyota (2005) was not required to recall Prius vehicles but a Toyota spokeswoman noted that a "software glitch in some 2004 and 2005 Priuses can make a warning light come on without cause, and in some cases shut down the gas engine altogether . . ."[73] They would notify owners, they promised.

Toyota (2010) admitted its Prius cars have an "anti-lock brake-system problem, which required a software change to fix . . ."[74]

Toyota (2014) recalled 2.1 million Prius vehicles because of a "software malfunction that can cause the engine to stop."[75]

Mechanical Problems Trigger Software Failures

All vehicles that rely on software *should* have failsafe measures that protect people from mechanical failures interfering with the secure functioning of the car's software, but they don't.

When mechanical failures cut power to the engine or stop an electronic control unit from working, certain safety features can become disabled – even though they're not supposed to.   For example, according to *The Wall Street Journal*, airbags are supposed to work even after power is lost.[76]

As well, steering and braking can become difficult or even impossible. In this list of some of the mechanical recalls that interfere with software electronics, I have included the GM ignition switch issue discussed more fully in a later chapter.

Chrysler (2014) recalled 350,000 vehicles over problems with the ignition switches that can "either become stuck or move without warning"[77]. In other words, the ignition switch can turn off the car while you're driving, which means that power is cut to the engine and that can cause the steering wheel and brakes to malfunction and the airbags not to deploy.

Chrysler (2014) recalled 907,000 trucks, SUVs and cars because a problem with the alternator can cause the engine to stall and "the electrical system to fail, as well as knock out power-assisted steering, antilock brakes and electronic stability control. It can even cause fire or smoke, according to documents Chrysler filed with the U.S. National Highway Traffic Safety Administration."[78]

GM (2014) recalled millions of vehicles due to a faulty ignition switch that would turn the vehicle off, cutting power to the engine, steering wheel and brakes and causing failure of airbags to deploy. *Reuters* reported in February 2014 that The Center for Auto Safety estimates at least 303 people have lost their lives due to this known problem. [79]

Although, by 2005, GM engineers had found a solution to fix this problem, a "business decision" was made *not* to fix it. GM continued to knowingly build cars with deadly ignition flaws for years[80] and this resulted in many families losing loved ones in terrifying crashes that were often put down to driver error.

GM (2014) recalled 2.44 million vehicles because the cars could have corrosion in the wiring harness for the body of a control module, "which could result in brake lamps failing to illuminate, or could cause brake lamps to light up when they're not supposed to.

"The condition also could disable cruise control, traction control, electronic stability control and panic braking assist . . [The company said] it's aware of 'several hundred complaints' and 13 crashes but no fatalities. It said it issued a technical service bulletin in 2008 and conducted a smaller safety campaign of 2005 model year vehicles in January 2009."[81]

Ford (2014) recalled "850,000 vehicles because an electrical glitch could prevent the air bags from deploying in a crash."[82]

Subaru (2013) recalled certain cars built between 2010 and 2013 because their remote engine transmitters "might have a loose internal battery clip that when dropped [that] sends a signal to start the car. If this happens *the engine can run for 15 minutes, turn off, and restart. This cycle could repeat until the car is out of gas.* In a garage this could lead to dangerous levels of carbon monoxide."[83] [Italics added.]

Volkswagen (2012) recalled 2,471 Beetles due to a faulty occupant control module in the front passenger seat. If the seat gets wet, "the module may malfunction and may not detect the presence of a child restraint. If that happens, the airbag will not shut off and will deploy in the event of a collision. Such a scenario could greatly increase the risk of serious child injury."[84] In this case, it is not a software coding flaw but a physical design flaw (allowing water leaks into the electronic module) that then causes software to malfunction.

Over the Air Software Updates

Tesla and BMW can fix a bug by changing your car's software "over the air". This means that you don't even have to bring your car to the dealer. Ford, GM, Toyota and other automakers are planning to do the same.

While this is an excellent idea, there are some key problems with this solution that need to be addressed across the industry:

- Without a sound, safety-first architecture, over the air software updates can open up further opportunities for hackers and
- Without explicit approval by the owner of the vehicle, the driver never knows when something has been "tampered with" or changed in the car.

Drivers should always be informed when their cars' software has changed but it is not clear to me at this time whether car companies update the software this way.

Also, can the software be changed *while* you're driving?

I hope not.

CHAPTER SEVEN

Toyota's Treachery

Treachery: Betrayal of trust; deceptive action or nature.[85]

To this day, despite the fact that their automobiles are basically giant computers on wheels, and a team of software engineers has found the code that can trigger SUAs in a Toyota Camry,[86] Toyota still insists that the many deaths and serious injuries due to their run-away cars are only the result of floor mats and sticky pedals.

Or driver error.

With at least one confirmed fatality due to software[87], hundreds of wrongful death lawsuits due to SUA and vehicles that continue to experience SUAs after their floor mats and sticky pedals have been fixed,[88] something is wrong and we will not be able to understand the full breadth of the software problems while automotive companies keep their software code top secret.

The reason I have put this chapter together is to help you understand that without a different set of regulations for software safety, and a way for "crowd sourcing" to capture what individuals are experiencing with faulty software, the car industry will easily be able to keep the public in the dark.

When reading through this Toyota saga, please remember that scores of people have died and many others have been injured for life.

Toyota's Twists and Turns

After all is said and done, here is what we so far know about Toyota's problems with Sudden Unintended Accelerations. Note that a large part of the facts are focused on floor mats and sticky pedals, but toward the end we roll back to what many have considered has also been the problem from the outset: faulty software.

Parts of this story were discussed in an earlier chapter but I have included some of the same items within this chapter's overall Toyota timeline to provide you with a full view of what happened.

As highlighted by notable U.S. officials, Toyota knowingly delivered cars that had deadly (physical) design flaws. Understanding how an automaker used its might to shirk its legal responsibilities and misinform the public, putting people's lives at risk, is important background when structuring a legal framework to monitor software safety.

It is a legal framework we desperately need.

In addition, certain facts about software safety that the Toyota engineers were concerned about with respect to SUAs did not come to light until a Japanese to English translator made the information public.

The Toyota engineers "sometimes admitted it was the electronic parts, the engine computer, the software, or interference by radio waves." -- Betsy Benjaminson, Toyota Whistleblower[89]

As you read through the summarized version of events, remember the enormous effort it took to take Toyota to task, effort supported by heavy pressure from the media, lawsuits, whistleblowers, government regulators and the U.S. Department of Justice.

And ten years.

In 2014 Toyota admitted they both broke the law and also lied and covered up deadly facts. Assistant Director of the FBI, George Venizelos, summed it up quite succinctly when he said:

"Toyota put sales over safety, and profit over principle . . . The disregard Toyota had for the safety of the public was outrageous. Not only did Toyota fail to recall cars with problem parts, they continued to manufacture new cars with the same parts they knew were deadly."[90]

In this chapter, I have included the summary of the Toyota story while at the back, in the Appendices, you will find additional supporting detail along with a full suite of footnotes and references. You will also find a sampling of some of the SUA-related complaints filed with the NHTSA by Toyota's customers.

Toyota SUA Timeline

As you read through the following, note that former NHTSA regulators hired by Toyota helped to end four of the eight NHTSA investigations into the company's SUA issues. All four of the "halted" investigations were related to problems with Toyota's electronic throttle systems.[91]

- 2001 – The 2002 Camry was substantially redesigned. According to the Safety Research & Strategies Inc., new or revised vehicle systems included:
  - the electronic throttle control system (ETCS-i) and
  - transmission and braking systems which consisted of "an accelerator pedal sensor, a throttle control motor, a throttle position sensor and the engine control module (ECM)."[92]

- In 2002, a Toyota Camry owner filed the first complaint with the NHTSA due to sudden unintended acceleration. Toyota issues "Technical Service Bulletin TSB EG017-02 to update the Electronic Control Module calibration to address 'engine surging' on 2002 Camrys . . . The Engine Control Module (ECM) calibration [was] revised to correct this condition." [93]

- In 2008, top engineers and management at Toyota in Japan were very concerned about their Imperial Family's "speed control" problems with their luxury Toyota. As reported in

The Japan Times, there were concerns about the heir to the throne's safety, and something had to be done.

"The problem seemed rooted in electronics — but its solution was elusive, even to all those trained minds. [For the Imperial Family] Toyota **replaced the gas pedal, the throttle system and the engine computer** at its own expense." [94] [Emphasis added.]

- Toyota also knew that their vehicles had floor mats and sticky pedals that could cause SUA[95] but:
  - Did not report this to the NHTSA (which is the law)
  - Continued to claim driver error, and
  - Continued to build and sell these defective vehicles for years, resulting in at least 89 deaths by mid-2010.

- In 2009, the company started a secret and "urgent" process of trying to fix the sticky pedal problem in the U.S., but after the very public and horrific crash of Mark Saylor and his family, *Toyota stopped the fix* and told their staff to deal with issues by telephone and not leave a paper trail concerning the sticky pedal problem.[96]

- In late 2009, Toyota began to recall vehicles to fix floor mats for some of the models they knew had problems, but not all of them. Then, in early 2010, it was clear that the floor mat problem was not fixing the SUA situation so Toyota recalled the vehicles a second time, this time to address the sticky pedal problem they'd already known about.[97]

- In May of 2010, NHTSA reported that 89 people died and 57 were injured between 2000 and mid-2010 as a result of Toyota's SUAs.[98] (More have died since then but I am unable to collect a full view of the publicly reported figures.)

- In August of 2010, referring to a report by the NHTSA, the U.S. Transportation Secretary Ray LaHood pronounced, "The verdict is in. There is no electronic-based cause for unintended high-speed acceleration in Toyotas. Period."[99] It is not clear why he claimed this because the detailed reports do not state this categorically.

- In 2011 NASA said that they couldn't find a problem with Toyota's software but pointed out that just because they couldn't find a problem, it didn't mean a problem didn't exist.[100]

- Ill-informed pundits immediately began to claim that the media "owed" Toyota an apology[101] because NASA couldn't find the problem with Toyota's software. These pundits didn't mention that NASA noted in its report that there *might* be a software problem, though.

- In 2012, Betsy Benjaminson, a translator for a law firm defending Toyota, began to notice that there were serious discrepancies between what Japanese engineers were considering as the problem – computer electronics – and what the PR messaging from Toyota was – floor mats and sticky pedals. After speaking with her Rabbi for guidance, she provided confidential documents to the media and the U.S. Congress.[102]

- In December of 2012, Toyota settled an economic loss class action suit for $1.6 Billion (suit claimed that multiple recalls on their vehicles had rendered the resale value of Toyotas to be lower than it otherwise should). Toyota continues to deal with hundreds of other lawsuits dealing with wrongful death and injury.

  This $1.6 billion settlement regarding "economic loss" is related to vehicles that are outfitted with electronic throttles and have undergone multiple recalls.[103] [104]

- Between 2010 and 2012, Toyota was fined a total of $66.4 million by the NHTSA for three separate breaches of the law.[105]

- In 2013 during the *Bookout v. Toyota* case in Oklahoma, Michael Barr, an embedded electronics engineer, proves that Toyota's software *can be the cause* of some of Toyota's SUAs – he finds the actual technology problem and arranges to have it tested with two Toyota vehicles. The jury finds Toyota guilty and Toyota settles, but Barr's

800-page report cannot be made public due to the terms of the settlement.[106]

- Toyota continues to claim that they have no SUA-related software problems, pointing to part of the 2011 NASA report.

- In 2014, Toyota finally admitted that it lied – for years – to the regulators and to the public about deadly flaws in their vehicles.[107]

- In 2014, the U.S. Department of Justice levies Toyota with a *criminal fine* of $1.2 billion (6.7% of its most recent year's profits) but agrees not to put any Toyota executive or employee in jail for breaking the law while the company sorts itself out over the next three years. According to *Reuters*, "the case was the first federal criminal case of its kind since the passage of the first U.S. auto safety law 48 years ago."[108]

SUA-Related Facts in General

One can rightly assume that software is likely not a factor in each and every one of Toyota's SUAs. But given the facts, below, it is not clear to me why Toyota still denies they may also have an SUA software problem:

- The reporting of SUAs is disproportionately higher in Toyota vehicles that are technologically more complex[109],
- Engineers at Toyota wrote of a "ghost in the machine" (which we testers call a mysterious type of bug because we can't easily reproduce it) and were so concerned about the computer electronics in the Japanese Imperial Family's Toyota that they refreshed the vehicle's software,[110]
- The Barr Group has pointed specifically at the line of code that causes SUA and *has tested and proved this* using two Toyota vehicles. The Barr Group also identified 80,000 lines of what we call in the business "spaghetti code". (Spaghetti code is messy software that doesn't follow software standards and is very often the source of a disproportionate percentage of bugs.)
- Toyota did not have a rigorous code development process that followed acceptable standards for safety critical software. The company only "sometimes" did reviews of *some* of the code. It also had no bug tracking system for the code reviews they undertook. This is problematic because 100% of the software should undergo a code review and a log of each bug found should be kept (to help identify areas of concern, patterns of defects and problem areas within the code).
- …and the team found a number of additional software bugs not related to SUA.

Each of these breaches leaves the public at risk. Even if you don't drive a Toyota, you could be hit by one when its software ignores the driver and races through a stop sign or slams into the back of your car as you sit at a red light.

Now, it is important to understand that software standards are not software laws. Individual companies have different software standards that they can choose to follow or not. Industries can have voluntary standards that companies like Toyota can choose to follow or not. However, when there are standards in place and companies choose not to follow them, or different vendors have a mix of standards making it impossible to align to one standard, making the software more prone to bugs.

Oh, and two more facts about Toyota's software situation:

- *SUAs are continuing to happen even after the floor mats and sticky pedals are fixed.*
- Right after the verdict in favor of Jean Bookout, Toyota entered into an "intensive settlement process" with "the hundreds of U.S. state and federal lawsuits alleging that defects caused its vehicles to accelerate suddenly and crash, resulting in serious injuries and deaths."[111]

Byron Stier, a professor at Southwestern Law School in Los Angeles, said, "The watershed moment [for Toyota] must have been that loss in Oklahoma . . . It was such a shock. People thought sudden acceleration was a dead issue, but the verdict changed everything."[112]

Well, almost everything.

As of April 2015, Toyota has not issued a single recall to fix the deadly software flaws Michael Barr and his team found in the Camry code.

Toyota Software Still a Concern

Toyota software continues to be a major concern for some customers. Robert Rugini, ironically an electronics engineer, owned a 2010 Toyota Corolla that was recalled for both the mat and sticky pedal issues. Even after his car had been retrofitted as a result of both recalls, his wife experienced three SUAs.

After each of the first two SUAs, he took it to the dealer but was told they could find nothing wrong with it.

The third SUA resulted in a crash.

In 2014, Mr. Rugini filed a complaint.

Mr. Rugini also reached out to Sean Kane, an automobile safety expert. Junko Yoshida of the *EE Times* reported that Kane indicated he is aware of 164 complaints that are very similar to Rugini's – floor mat and sticky pedals have been fixed; yet the vehicle still suddenly accelerates on its own.[113]

Mr. Rugini has filed a complaint with the independent lawyer assigned to oversee the 2014 Toyota settlement with the U.S. Justice Department, claiming that "Toyota Motor North America may already have broken the terms of the March 2014 deferred prosecution agreement by making misleading statements and concealing information on a safety issue related to unintended acceleration" and has asked that his case be investigated.

What is "good enough"?

In closing out the part of this chapter focused on Toyota, I would like to highlight that in February 2015, J. D. Power announced that Lexus owners reported a high level of satisfaction with the car's dependability. This should not side track our software safety discussion or the recommendations in this book.

When Cars Decide to Kill was not written to refute Lexus owner satisfaction with their vehicles. It was written to address the lack of software safety laws. We should remember that:

- A car can be dependable 99.999% of the time. It's the other ".001%" that you have to worry about.

You can have a highly dependable car that is highly reliable most of the time but it just takes one SUA to kill you.

Toyota continues to move ahead with all sorts of new technologies that are, as of this date, not being pro-actively monitored for quality by a third party. When it came to the company's floor mat and sticky pedal problem, the U.S. federal regulator was unable to "catch" Toyota breaking the law for ten years.

Who is going to oversee Toyota's newest "basket of technologies" and wouldn't it be better to catch problems before they are delivered to customers, rather than after?

> *"Toyota has introduced a basket of technologies that bring the possibility of autonomous driving that bit closer, including a 360-degree view parking assistance system, a pedestrian braking system, an advanced form of cruise control (using wireless communication between vehicles) and **a more accurate** form of automatic lane-keeping."[114] [Emphasis added]*

A more accurate form of automatic lane-keeping?
More accurate than what, and who says?

Other Car Manufacturers' SUAs

Toyota is not the only automaker that has trouble with sudden unintended acceleration. A search of the U.S. SafeCar.gov complaints database in February 2015 found there were over 1,700 complaints related to vehicles suddenly accelerating with 41% from Toyota owners.[115] These complaints are, of course, from those drivers who survived. It is uncertain how many of these situations happen around the world, and how many people have either died or been wrongfully convicted of dangerous driving as a result of the software taking control of their vehicles.

Many complaints were laid against Toyota, Ford and GM products, but there were complaints against other makes of vehicles as well – BMW, Chrysler, Honda, Hyundai, Kia, Land Rover, Mazda, Mercedes-Benz, Mitsubishi, VW, Volvo. I have provided some examples in an Appendix.

…and then Google Hires a Regulator

Although the National Academy of Scientists, after reviewing the NHTSA's own inquiry into the Toyota SUA crisis, reported in 2012 that the *"federal regulators had lacked the expertise to monitor electronic controls in automobiles"[116]*, that same year Google hired the Assistant Deputy Director of the NHTSA to be their Director of Safety for Self-Driving Cars.[117]

Does it seem reasonable that a senior regulator of the very agency that "lacked the expertise" to monitor software safety in today's cars is well positioned to be in charge of safety for tomorrow's self-driving cars?

Or is something else going on?

CHAPTER EIGHT

And Then There's GM

The death toll from the GM ignition switch defect has risen to 80 as of April 2015. The final number of people who have died as a result of the faulty GM ignition switch is expected to be 148.[118]

I would be remiss not to give GM a chapter of its own.

It is important to remember that the ignition flaw discussed here is a mechanical failure that shuts power to the engine, brakes and airbag deployment. It is not a software failure, per se.

The reason I am providing a brief timeline on GM's breach of U.S. law and public trust with respect to a mechanical, not software, failure is to show how very important it is that we *not* rest on the existing automobile safety and regulatory processes that have failed so tragically.

As you can see from the Toyota saga and the GM timeline, two carmakers have acknowledged that they chose to put human life at risk knowing that some of their customers would endure a terrifying death.

The truth is, their decisions actually put everyone at risk because a number of the crashes have involved vehicles that were not Toyota or GM.

The GM timeline I have summarized here follows just one GM vehicle and one driver, Brooke Melton, who died on her 29th birthday because her ignition switch turned the engine off while she was driving. It was a rainy night and her vehicle hydroplaned, crashing into a river.

In my opinion, she died because GM's board, executive and management teams didn't do their jobs. Their jobs should include ensuring that GM produced the safest possible product, following through on processes where equipment failures and dangers to customers are properly reported and fixed. The culture of apathy within GM is staggering. There are reports of meetings where management was warned (for years) of the ignition problem and no one bothered to do anything at all about it. Worse, quality experts were afraid to come forward because they feared for their careers.

> *There was the "'GM nod'—when everyone in a meeting agrees that something should happen, and no one actually does it. ...[A] GM safety inspector named Steven Oakley is quoted telling investigators that he was too afraid to insist on safety concerns with the Cobalt after seeing his predecessor 'pushed out of the job for doing just that.'"*[119] --
Businessweek

As a result of Brook Melton's parents' determination to hold GM accountable for her death, GM's Board has established a special risk management committee and the CEO, Mary Barra, is working to establish a culture that fosters the proper warning systems and communication paths to executives.[120] As well, the

U.S. Department of Justice is investigating whether the company is criminally liable for having hidden defects that killed dozens and dozens of people from both their customers and federal regulators.

GM's Faulty Ignition Switch Timeline

- 2002 – faulty ignition switches start to be installed in millions of GM cars,
- 2004 – a GM engineer finds the problem on the 2005 Chevy Cobalt while testing it before release,
- 2004 – GM releases Chevy Cobalt with the known ignition problem,
- 2005 – Brooke Melton, a pediatric nurse, purchases the Chevy Cobolt,
- 2010 – The engine shuts down while driving, so Brooke Melton brings her 2005 Chevy Cobalt into the dealer to have it looked into. Her hand-written notes state "key locking in ignition" and "suddenly shutting off while driving and unable to turn vehicle". The dealer assures her that all is well with the car.
- 2010 – The day after picking up her 2005 Chevy Cobalt from the service dock at the dealer and being told nothing was wrong with her ignition, her engine shuts down while she is driving at 58 m.p.h. in the rain. This turns off power to the brakes and airbags. It also interferes with her seatbelt. She tries to brake, fishtails, hydroplanes, and then
  . . .

*. . . another vehicle "plowed into the passenger side of Melton's Cobalt: 3,000 pounds of steel, glass, plastic, and human smashing into 3,000 pounds of steel, glass, plastic, and human. While Brooke's lap belt glued her waist to the seat, **her shoulder harness went slack the instant the engine shut off**. As the side of her car caved in, Melton's torso, neck, and head whipped violently to the right, the force equivalent to falling from the 16th floor of a building."[12.]* [Emphasis added.]

- 2010 – Brooke's car ends up in a creek. She is found sitting at the wheel of her vehicle with water up to her shoulders and is rushed to hospital. She dies before her parents can get to the hospital. <u>Brooke's parents file a lawsuit against GM for wrongful death</u>.
- 2013 – The Meltons are awarded $5 million and their lawsuit ultimately triggers the recall of 29 million GM vehicles.

The Melton case "exposed that GM knew about the faulty switches for more than a decade but failed to recall the cars until this year [2014]. It also touched off a crisis at the company that has resulted in 54 recalls involving 29 million vehicles this year. And it brought federal investigations, coverup allegations and a $35 million fine from federal regulators for delays in reporting safety problems." [122] (Associated Press)

- 2014 – An NHTSA investigation stalls as GM delays providing answers to the 107 questions presented to them. In April, the NHTSA fines GM $28,000 "for failing to answer questions about its ignition switch recall."[123]

- 2014 – Although Brooke's parents are in a "constant state of grief" [124], they believe that GM withheld evidence from the original trial and ask the courts to re-open the case to secure a jury trial. To do this, they will give back the $5 million settlement.[125] "Their goal in this lawsuit is to uncover what GM knew about the defect and about the hundreds of deaths and injuries that were caused by it."[126]
- 2014 – The U.S. Federal regulator, NHTSA, issues against GM the maximum fine permitted by law: $35 million. This represents less than 1% of GM's 3rd quarter profits for that year.
- 2014 – The U.S. Justice Department opens a criminal investigation against GM.
- 2014 – GM pays for a study known as the Valukus Report to look into the ignition switch issues. "Valukas chalked it up to incompetence." The Meltons lawyers claim in was a cover up.[127]
- 2014 – GM dismisses 15 employees, including at least eight executives, after an internal investigation found "a pattern of 'incompetence and neglect' that led to 11 years of delays in recalling millions of cars for a fatal defect."[128]
- 2014 – GM assigns lawyer Kenneth R. Feinberg to oversee its GM Ignition Compensation Claims Resolution Program
- 2015 – GM settles with the Melton family for the wrongful death of their daughter. "Neither GM nor lawyers for the family of Brooke Melton disclosed terms of the agreement,"[129] reported *Agence France Press*. However, the lawyers representing the Meltons pointed out that as a result of their second lawsuit, they were able to secure

"documents produced under seal for the suit [that] show GM management and engineers knew the switch raised safety issues and ignored the problem." These documents will be able to be used by others suing GM.

Although we – yet again – have a secret deal between a powerful automaker and a grieving family, the Meltons' two lawsuits triggered a process to help the families of other victims because:

- Documents have been produced during the second trial that can be used by other victims and
- GM set up the ignition switch Compensation Program in 2013.

At the time the program was set up, GM acknowledged only 13 deaths associated with their business decision not to fix the faulty ignition switch (and to keep building new cars with the faulty switch installed, remember).

The GM Compensation Program is ongoing but has grown to confirm that the ignition switch problem has killed 80 people – and this number will likely rise to 148 deaths.[130] The numbers have been rising each week as the program continues to evaluate the claims. (Those who accept payment through this program give up their right to sue GM.)[131] There are an additional 104 wrongful death lawsuits against GM at this time.[132]

- In July of 2014, lawmakers and lawyers estimated 100 deaths[133],
- The Center for Auto Safety claims the number is 303 [134].

Why the discrepancies?

That's another thing that needs fixing. Helping the public – and the media – actually make sense of it all and know when the facts are being reported rather than massaged or masked in corporate press releases would be a helpful start.

Pre- and Post-Bankruptcy GM

In 2009, U.S. Bankruptcy Judge Robert Gerber told GM that it "didn't have to worry about lawsuits over cars made before its $49.5 billion government bailout." The judge later "expressed doubts about his decision", especially in light of the faulty ignition switch situation.[135]

Judge Gerber is now deciding how to "fix" his 2009 ruling and expects he will come up with an answer in the spring of 2015.

CHAPTER NINE

And Audi, Too

This chapter is not on the subterfuge of an automaker. Audi behaved responsibly and responded appropriately to the flaws identified in their vehicles.

Rather, this chapter is on the relentless string of ill-informed media reports that have either heavily relied on press releases or unwittingly kept reporting the wrong information, misleading the public. I have chosen to use Audi as an example for two reasons:

- Because many media reports dove-tailed stories about Audi's fairly straight-forward investigation regarding SUA with Toyota's convoluted and ultimately criminal investigation and

- Even after decades as a software test expert, after three months of conducting research for this book and after two months of intense, focused writing, I still got hoodwinked by what I'd been reading in the media and temporarily missed the facts.

During a review of a draft of this book, two people reminded me not to forget the Audi SUA problem that occurred in the late 1980s.

"Not applicable to technology issues," I replied. "It was only a pedal problem and Audi was exonerated."

Exonerated. It's not a word I normally use, especially not in my profession.

Plus, how did these book reviewers so clearly remember something that had happened back in the 1980s?

So, let's look at the "exonerated" question first. In his blog, Michael Barr summarized the Audi NHTSA report on the sudden acceleration problems in the 1980s. He highlighted a key finding quoted from the report itself:

> *"Some versions of Audi **idle-stabilization system were prone to defects** which resulted in excessive idle speeds and brief unanticipated accelerations of up to 0.3g. These accelerations could not be the sole cause of [long-duration unintended acceleration incidents], but might have triggered some [of the long-duration incidents] by startling the driver."* -- NHTSA's official Audi 5000 report [Emphasis added.]

In short, the Audi SUAs, in some cases, were likely technology flaws that startled the driver and triggered human error. Audi was not fully free from blame at all. It was ***not*** exonerated.

I ran an Internet search for "Audi exonerated" and found an explanation as to why two of my book reviewers so easily remembered this situation from the 1980s.

After the Mark Saylor Lexus crash in 2009, the media put more focus on Toyota – and, as part of the story, often claimed the 1980s NHTSA's investigation "exonerated" Audi of any sudden acceleration issues. At times, articles either inferred or outright claimed that because Audi had been "exonerated," Toyota was innocent, too.

The FACTS

When Audi's SUA issues were brought to the media's attention in the late 1980s, Audi behaved like a responsible corporate citizen. It responded to the findings of the NHTSA by adding an automatic transmission interlock that made it impossible to shift into drive or reverse without a foot on the brake.

In my opinion, the media behaved less responsibly. As Michael Barr pointed out, "there appears to be a fundamental contradiction between the way that Audi's problems are remembered now and what NHTSA had to say officially at the time."[136]

The FICTION

Probably the most startling example of fiction reported as fact about the 1980s Audi SUA appeared in *The Atlantic* in the summer of 2010. Under the bold headline, "NHTSA: No, Toyotas Don't Accelerate Unless You Press the Accelerator," *The Atlantic* writer heavily quotes P.J. O'Rourke's severely misinformed opinion:

> NHTSA's Audi report presented evidence that "proved what everybody who understands how to open the hood of a car had known all along about [SUAs]: 'Pedal misapplications are the likely cause of these

incidents.' Yes, the dumb buggers stepped on the gas instead of the brake." [137]

The writer at *The Atlantic* goes on to say, "Naturally, when Toyota's troubles started, I immediately thought of this passage [from O'Rourke's book]. I decided to poke around in the data about the accidents, and found that this time around, they sure did look a lot like, um, pedal misapplication." [138]

In 2011, *The Atlantic* ran a piece on how safe Toyota vehicles are and referred to the NASA report again. Interestingly, they didn't run an article in 2014 on Toyota's criminal charges and admission that it misled U.S. regulators and the public.

Wall Street Journal, March 2010 – In an article headlined "Audi Case Sets Template for Toyota's Troubles," the writers stated that "Audi . . . was ultimately *exonerated* of building defective cars, but not before its sales and reputation took a pounding at a strategically critical moment."[139]

New York Times, March 2003 – "It took Audi 13 years to rebuild its sales in the United States after the sudden-acceleration fiasco, even though the government *exonerated* the company in its 1989 study."[140]

The New York Times, May 1997 – "No matter that Audi was subsequently *exonerated*."[141]

Law360.com, February 2015 – "Audi AG . . . was ultimately *exonerated* of building defective cars, but not before sales of all Audi vehicles — not just the 5000 model — fell by 84 percent."[142]

Note the date of the last quote: February 2015. This particular Law360.com blog was written by lawyers, not reporters, who appear to defend automakers.

Final Word: Audi Did the Right Thing

But the company was *not* "exonerated."

After so many misleading and just plain inaccurate excerpts, I feel the need to repeat: **the NHTSA did not exonerate Audi**.

Audi did take its corporate responsibility seriously and fixed *its* part of the problem, which then reduced the likelihood of startling the driver because of an unintended acceleration in one of their cars.

CHAPTER TEN

The Hack Factor

"[Your] car is not a simple machine of glass and steel but a hackable network of computers." – Forbes Magazine[143]

What can you do if, as you're driving along, someone wirelessly hacks into your car's software, takes control of your vehicle, swivels the steering wheel so that it swerves into the next lane, honks your horn over and over again, causes the car to speed up and drive straight through a red light, and then suddenly brake?[144]

The answer?

Nothing.

Although – as of this moment – I can find only one reported incident of malicious automobile hacking (involving over 100 cars in Texas), it does not mean it's not secretly happening already but we just don't know about it. Remember, it took dead bodies, dozens of lawsuits, a NASA investigation and Michael Barr's expert examination of software code *plus* ten years for Toyota's software problem with Sudden Unintended Acceleration to come to the light of day.

At this point in time, *no one can be sure that hacking into cars is not happening today.* A U.S. Senator's report highlighted in early 2015 that most automobile manufacturers have acknowledged they are "unaware of or unable to report on past hacking incidents."[145]

The first publicly reported case of malicious automobile hacking took place in 2010 around the same time the U.S. Transportation Research Board (TRB) commissioned a report on the implications of the widespread use of on-board electronics.[146] While the computer scientists funded by the TRB were busy proving that they could hack into and control a car from a distance of 1,000 miles, a disgruntled, laid off car dealership employee sitting at home hacked into his former employer's GPS system and disabled its customers' cars' ignition systems and caused the horns to honk non-stop.

As reported in *Wired Magazine* online, "[The] Texas Auto Center began fielding complaints from baffled customers the last week in February [2010], many of whom wound up missing work, calling tow trucks or disconnecting their batteries to stop the honking. The troubles stopped five days later . . ."[147]

Who paid for the tow trucks? Who paid for the lost wages? In many industries, the cost of the fallout of software gone awry is almost never tallied up, a large chunk of it resting with the consumer.

There are hackers, and then there are helpful hackers. For over five years, the helpful hackers have been trying to get people to understand that *our automobiles are vulnerable to cyber attack*. As Pablos Holman (a helpful hacker) pointed out in a *TED-X Talk* "Your car is now a PC . . . When that happens, you inherit all the security properties and problems of PCs. And we have a lot of them."[148]

Here is a list of the parts of your car that can be hacked:

- Steering
- Brakes
- Gas Pedal
- Tire Pressure
- Odometer
- Speedometer
- Seat Belts
- Doors
- Trunk
- Airbags
- Possibly more . . . such as navigation

In the February 2015 report issued by U.S. Senator Ed Markey, it was noted "[s]ecurity measures to prevent remote access to vehicle electronics are inconsistent and haphazard across all automobile manufacturers, and many manufacturers did not seem to understand the questions posed . . ."[149]

Haphazard software design that does not follow a consistent structure and set of standards is a gold mine for hackers. Hackers can gain control using such features as Bluetooth, OnStar™, ConnectDirect™ and other systems that connect in cyberspace. Hackers can also infect a vehicle's software with a simple CD that contains malicious code designed to help the hackers do their dirty work.

Hacking as a Design Feature

In many newer model cars, the auto industry has installed software to help car dealers and navigation system providers work together to track and remotely disable vehicles if drivers do not keep up with their payments. That's why it was so easy for the Texas hacker to use his former employers' computer systems to take control of their customers' vehicles. While some of the systems designed to remotely disable your car can "hack" into it and disable your car only while parked, others can disable your car while you are driving on the road.

Either way, in an emergency situation, allowing someone else – even someone to whom you owe money – to decide if and when you can drive your car can put not only your life but the lives of those you love at risk.[150] What if someone is injured? Sick? Pregnant and needs to go to the hospital? What if you stopped at a red light in a dangerous part of town with three little children in the back seat?

This "financial security" feature is designed for the global car industry, not for you, and is totally unregulated.

Tracking Drivers Wherever They Go

In addition to the serious issues associated with the fact that nearly 100 percent of the vehicles on the road today are vulnerable to being hacked, U.S. Senator Ed Markey noted in his report "Tracking & Hacking," that there was "an aalarmingly inconsistent and incomplete state of industry security and privacy practices"[151] which put drivers at risk. The information collected by automotive companies about a driver's personal driving habits are often given by the carmaker to a third party for an indefinite period of time and with little to no ability for the consumer to opt out. According to his report, the types of information currently being gathered and shared by automakers include:

- Geographic location such as
    - Physical location recorded at regular intervals
    - Previous destinations entered into navigation system
    - Last location parked
- System settings for event data recorder (EDR) devices which can include:
    - Potential crash events, such as sudden changes in speed
    - Status of steering angle, brake application, seat belt use, and air bag deployment
- Fault/error codes in electronic systems
- Operational data such as:
    - Vehicle speed
    - Direction/heading of travel
    - Distances and times traveled
    - Average fuel economy/consumption
    - Status of power windows, doors, and locks
    - Tire pressure
    - Fuel level
    - Tachometer reading (engine RPM gauge)
    - Odometer reading
- Mileage since last oil change
- Battery health
- Coolant temperature
- Engine status

- Exterior temperature and pressure

Don't Forget the Car's "Black Box"

In 2012, *Computerworld* reported that the "black boxes" U.S. federal regulators now require by law (as of 2013) "might turn into the star witness to testify against you [depending on the jurisdiction in which the accident happened]. You may not think about or be aware of your vehicle's event data recorder (EDR), yet it is constantly recording evidence like a plane's 'black box' and is being used after a crash to explain why it happened."[152]

Various jurisdictions have different rules about whether or not the black box in your car can be used in a court of law. Your car could be registered in a jurisdiction that disallows the use of the black box but get in an accident in a jurisdiction that permits data from your black box to help determine fault, and you could be in trouble.

Now, remember – like all automakers these days – Toyota collects this type of information in its "black box" and experts found that the company had a software error in the black box, itself. Also, Toyota announced that it had a software error in the laptop that reads the crash data from black box. When companies collect such a breadth of information that can be used by them to refute customer service claims and liability in the event of a crash, yet the information can be inaccurate because of mysterious software flaws, drivers have nowhere to turn for support or relief.

Other carmakers likely have bugs in their black boxes, too. The trouble with the black box situation is:

- Regulators and courts have no idea at any point if the software capturing, reporting and interpreting our car's black box data is sound, as code can change from week to week and year to year, and
- As U.S. Senator Ed Markey pointed out in his report "Tracking and Hacking", each of us has a right to know what the data is being used for and should have the explicit right to opt out of certain forms of information gathering without being "punished" by the manufacturer and losing key features such as navigation. In short, why should I be required to accept having a company gather and share my personal information (about where I drive and when) just so that I can use the navigation device in the vehicle?

And, by the way: Why did the U.S. regulators define the data that must be collected without being very, very clear about how it can be used legally?

Track, Hack … What the *HECK*!?@*?!

Today's automobiles do not have the required security features to prevent **someone else** from taking control of your car or knowing all about what you do and where you go. Someone else a thousand miles away can now take control of your car and drive you through a red light, into a brick wall, over a cliff or up onto a curb and into a crowd of people. If that happened, would you be alive to explain that it wasn't actually you who made the car go out of control but a "mystery person" somewhere on the planet?

And who would believe you, anyway?

In 2013, *Forbes Magazine* writer, Andy Greenberg, who covers "the worlds of data security, privacy and hacker culture", reported what it felt like when he drove a car that had been hacked by a couple of "helpful hackers":

> *"Stomping on the brakes of a 3,500-pound Ford Escape that refuses to stop–or even slow down–produces a unique feeling of anxiety. In this case it also produces a deep groaning sound, like an angry water buffalo bellowing somewhere under the SUV's chassis. The more I pound the pedal, the louder the groan gets–along with the delighted cackling of the two hackers sitting behind me in the backseat."[153]*

At least Andy Greenberg knew that there were two helpful hackers doing this to him, and he was driving at 5 mph on a small road with no traffic.

CHAPTER ELEVEN

The Big Guys Need to Get It

Nobody can test quality into a product. ***Quality must be built*** *into a product and that takes the whole company, including "The Big Guys".*

Although there are no software safety standards for car companies, other industries – such as the military, medical equipment and aerospace industries – must follow strict software development processes or face consequences. Even in those industries that produce exceptionally robust software, mistakes happen but they happen with less frequency.

Whether you're baking baguettes or building software to run a factory or hoping to put cars with no steering wheels on the road, the principles of quality management across all industries are the same. You just apply them with greater attention to detail and more regulatory oversight if the software being developed could hurt a human.

No matter how much of a whiz-bang a software engineer is and no matter how excellent your quality management experts are, they can't build and test safer software without an organization that supports quality. To do that, the organization must be structured with the understanding that you don't test quality into a product, you build it in.

Corporate culture cannot change without engaged leadership from the company's executives and board members. Yes, the most important aspect to software quality is this: The Big Guys Need to Get It. All too often, quality management is a mystery to the people running our multi-national corporations and, stunningly, even to those in charge of large IT Departments.[154]

For example, the Y2K situation was real. The reason the world didn't turn upside down when the Year 2000 "hit" is because in the years leading up to 2000 executives around the world paid attention. The Y2K situation was a mostly "non-event" because the world's executives did a great job of focusing their energy on the problem. Even though some companies, like Visa, wisely used the lessons they learned during Y2K to improve their testing processes[155], other companies fell back to their old ways and even punished people who had done a good job in the quality management department.

One of my professional mentors had become the Chief Information Officer for a technologically complex insurance company with a global reach. His mandate was to have his IT department work its way through all of the legacy (old) software and other technologies to make sure that the insurance calculations would be both accurate and not breach any laws (insurance is a regulated industry). A tremendous, focused effort was applied.

Thousands upon thousands of problems were found and fixed in advance of the Year 2000. Everything went off without a hitch and when the Year 2000 arrived, the company's technology functioned extremely smoothly.

The insurance company's Board invited my former mentor to a board meeting in early 2000 where he was able to give the good news.

He was fired.

Why?

Because, as the Board said, "We spent so much money and there weren't even any problems."

Well, that *was* the point. You're supposed to find and fix problems in the test world, not the real world. This is why the motto for my company, Test Matters, is this:

The sound of a good test is silence.

Customers don't call to complain, no urgent meetings are held to try and manage the PR nightmare and there's nothing much to do except get on with business as usual.

How MBA-Based Ideologies Hurt Us All

Technology is the backbone of most multi-nationals and yet in the last fifteen years, just as it was becoming critical to have technology-savvy experts guide a company's decision-making processes, organizations have turned almost exclusively to those with MBAs to act as leaders. Some companies require that even before an employee will be considered for an executive role, he or she must hold an MBA. In fact, I have seen job opportunities for senior Directors of Quality Assurance – and the educational requirement is "MBA."

This makes no sense.

The strength of any team is in the various skills each player brings to the table. Insisting that everyone be a quarterback means that nobody's running defense. Similarly, insisting that everyone hold an MBA means that a company is asking the entire executive to speak and think the same way. In my opinion, MBA schools pay too little attention to the subject of technology quality fundamentals, so few executives really understand the impact their decisions have on whether or not their companies are delivering sound software.

It takes a longer amount of *time* to build safe software and it involves more than just programmers and testers. It involves the entire organization, led by the most senior executive.

Yet many executives fuss quite a bit to keep the "time to market" factor as short as possible and the software development costs as skinny as possible. In the late 1990s, car companies started to contract out a fair chunk of the work they used to do within their own organizations – a situation that has caused them lots of problems that they are now trying to fix.[156] Why? Not only can millions of lines of software code add to complexity, but a patchwork of many suppliers' products increases complexity, too.

As Toyota once understood twenty-five years ago, corporate culture needs to support the delivery of a quality product. These days, a unified quality culture must somehow extend to all of the suppliers and contractors. Few companies have figured out a way to do that. Recognizing the hidden costs of seeking cheap suppliers, car companies are now re-thinking their outsourcing and engineering contract strategies.

When it comes to our cars, quality must take priority over the MBA-inspired priorities of outsourcing, lowering costs, gaining "synergies" and chasing an ever-increasing piece of the market share pie. Not that any of these things are bad, per se, but when companies are in the business of producing things that can kill people if something goes wrong, safety must be the number one priority at all times. This costs money and takes time, and executives need to understand and support this.

This isn't based only on more than twenty-five years of experience in testing and quality management[157], it is also now the considered opinion of both the President of Toyota and the Chief Executive Officer of Honda.

Akio Toyoda, the President of Toyota and grandson of its founder, said that the turning point for his company was 2003 when it changed its business strategy, moving the company's focus *away from quality* to increasing market share. This new strategy shifted corporate leadership attention from caring about the company's product and its customers to caring about how financially powerful it could become.[158] Once this happened, employee promotions were skewed toward those whose work supported the new strategy of making a lot more money, not those who strengthened the company's ability to deliver cars that didn't kill people.

Likewise, Honda's Chief Executive Officer, Takanobu Ito, announced in February 2015 that the company "would no longer pursue business expansion as its main target" because "the pressure to reach the objective contributed to the quality lapses that led to record recalls" in 2014.[159]   Earlier in 2014, Honda had announced that its profits had risen 7% based in part on cost cuts.[160]

But as Honda and other companies discovered in 2014 when 64 million recall notices were issued, cost cuts can be costly.

Quality isn't cheap but it pays off in the long run.

Supporting Executives with Quality Vision

The extent to which a company focuses on quality varies not only company-by-company but often department-by-department within each company because, generally, there are no tightly monitored rules supported by executives and there are certainly no laws. Sometimes, there are "voluntary guidelines" but there are no rules.

At the end of the day, though, executives can and do put pressure on the entire software development process and squeeze out key quality management activities to shorten that ever so important time-to-market factor.

In a world where there are no software safety regulations:

• Executives who want their teams to take the right amount of time to safely build and test their critical products actually put their own careers and companies at risk because (see next bullet)

• Other executives reap rewards by jumping the gun, encouraging shortcuts and getting new, exciting, and sometimes deadly, products out the door faster . . . and consumers lap it up, blissfully unaware that they are putting themselves and their families at risk.

If they think about it at all, many people hop into cars assuming that someone, somewhere is watching the carmakers to make sure they are building safe software.

Nobody's watching. Nobody's mandated to watch.

I have met many executives who actually very much want to deliver quality and even if they don't understand all the moving parts associated with quality management, their desire to serve their customers properly drives positive change. But it is quite a challenge for them and takes more effort than it should. They meet resistance every step of the way from other executives who are focused on cost cutting and worried about time-to-market.

To strengthen an organization's concern for the safety of its customers, responsible executives should be supported "from the outside" through mandatory software regulations that directly influence proper corporate behavior "on the inside".

Clear rules will help level the playing field and help those executives who want to do the right thing be able to do the right thing.

Chief Quality Officers?

You can tell a lot about a corporation's attitude about quality by looking at where the most senior QA/Test expert sits in the organizational hierarchy. Usually the answer is "the bottom of the management food chain."

This means that executives face another problem when trying to help the organization deal with quality: they do not hear the bald truth about product quality issues. Instead, the message gets filtered through many layers of management until very senior management hears a much rosier picture than what the test expert knows to be true.

Fear for your own job, fear for passing bad news upwards, fear of embarrassing your boss are just some of the worries that people have, exacerbated by organizational pressures that work against sound software creation, quality management and testing activities.

This is slowly changing in the car industry.

Toyota again understands the importance of direct executive oversight on Quality Management. As a result of the 2014 $1.2 billion settlement with the U.S. Federal Justice Department, the company has appointed "Chief Quality Officers for North America and other principal regions – *all of whom have direct lines* to President Akio Toyoda.[161]

In short, Toyota now has quality management executives who each have the ear of "The Really Big Guy". They've organized themselves for quality.

The rest of the auto industry has to do the same.

CHAPTER TWELVE

Getting Safer Software

> *"The highest level of safety is to submit to third-party checking. Unfortunately, the auto industry has escaped this safety-certification system . . . Basically, consumers are being asked to trust the automakers with their lives."* [162]
> -- Betsy Benjaminson, Toyota Whistleblower

The foundation of a quality software product is its architecture. I know I've said this before but it bears repeating again: Without a system architecture based on safety and security requirements, we cannot have safe cars.

Architecture is important because a badly designed car is a deadly car. For example, when Michael Barr studied the Camry software, he found:

- That the brake override that is supposed to "save the day when there is an unintended acceleration" occurs in a group of code that could "die" from memory corruption, thereby disabling many of the fail safes embedded in that code and
- Toyota used "an operating system in which there is no protection against hardware or software faults."[163]

What this means is that the safety override braking feature was written *inside* the program that could fail due to something called a memory leak. If a program shuts down because of a memory leak, all of the code in that program stops working – including any safety features tucked inside of it.

Imagine you had a spare key for a closet where you keep your valuables and then you lock yourself out of it . . . only to remember that you keep your spare key in the locked closet. That's basically what Toyota did. You can't expect a safety feature to protect you if it is put *inside* the system that it's supposed to protect you from. If the system fails, the safety feature inside it fails, too.

So in the above example, a proper, *safety-based* architectural framework would ensure that safety overrides sit "outside" the code that can potentially fail. As reported by Junko Yoshida in the *EE Times*:

> *"New findings include defective software that contains bugs, and -- in the 2005 Camry -- an electronic throttle control system with inadequate safety architecture, whose design created a single point of failure with no redundancy in place."[164]* – EE Times

How many cars on the roads today have a well-structured software design vs a poorly designed architecture?

We don't know.

Both helpful hackers and U.S. Senator Ed Markey have pointed out that the systems in our vehicles today are not standardized. Here are just a few reasons why we need to be concerned about this:

- With up to 20 million lines of code in any given vehicle, and with systems built by scores of car manufacturers and their suppliers, there are too many opportunities for errors and security breaches.
- Additionally, when a car decides to kill and the police start their investigations, non-standardized architecture makes it exceptionally difficult to find out what was really going on, especially because carmakers are very protective of their proprietary software.
- Without software standardization for the auto industry, it makes it very difficult for any regulatory agency to collect and interpret data (from car crashes and software-related complaints) that would help make improvements to software safety practices.

You can't just layer good software over bad software and hope it fixes the structural problems in the technology. It won't.

Worse, layering new code over top of old software can be even more dangerous. Why? Because dormant bugs often sit quietly in old software and these bugs can (and do) suddenly "come to life" when they meet up with the new code, with devastating consequences.

Besides, the old software was not coded to the correct safety-focused architecture we need for the 21st century.

How do I know that? Because we haven't democratically agreed upon what that should be. We haven't put our best minds to the question. We've just let companies individually come up with their own ideas without going back to the drawing board to architect a safety-first software design that will protect all of us.

That was fine in the "old days" (all of fifteen years ago) but as we look to the future, a future where automobiles will be making a lot more decisions than that little pony did when deciding whether to cross the road or not, leaving the security standards to each individual automobile company is not only dangerous, it's deadly.

And with "self-driving" cars in our future, and each carmaker relying on their own proprietary software, all of us should be very nervous, indeed.

We have a precedent for standardizing new technology at the turn of the last century. When electricity was introduced into homes, it was extremely dangerous. People came up with all sorts of weird (and deadly) products like electric tablecloths and there were no safety standards. Exposed wires were routinely placed in walls and each town and village in England, for example, had different voltages.

Finally, various regions came up with mandatory standards which is why, for example, when traveling overseas you must bring along the right set of plugs for that region.

And After the Crash . . .

There is another problem with "variations on a theme" when it comes to the electronics in our cars: it presents dangerous situations when First Responders (e.g., firefighters, police officers, ambulance attendants) arrive at the scene of a serious car crash. Why? Because if they need to use the "jaws of life" to cut through metal to try to save someone, without structural information about the car, they may be cutting through an energized circuit or an explosive device, putting lives at risk. To avoid that, First

Responders must know the year, make and model of the car and then look up structural information in a manual.

Recently, Mercedes Benz came out with a new way to get this critical information at the site of an accident. It's something they call Rescue Assist[SM] and here's the wonderful thing about that:

> *Mercedes Benz has **waived the patent rights** on the technology for Rescue Assist[SM] so that other car companies can reproduce it for their own vehicles, thereby making the rescue process consistent.[165]* [Emphasis added.]

This level of corporate care for the public and for our First Responders is to be commended.

And we need more of this kind of consistency across all automakers in the foundational software design of our cars as well as ongoing automotive cooperation if we want to keep the roads safe.

Getting the Right Design

As a society, we need a sound and standardized foundational design for the robot cars that will soon be driving on our roads, a design that meets a suite of both safety and security requirements that automakers and software companies must abide by. To do this, we need:

- A design that is safety-focused,
- Technology development standards and
- A proper regulatory framework to support these standards.

When it comes to design, look at the difference between Microsoft and Apple's operating systems and their security levels. Apple spent a great deal of time on a systems architecture that not only supports the seamless integration of current and future products (as well as user friendliness), but its operating system is also far and away more secure than the Microsoft Windows operating system.

In short, Apple spent a long time thinking about and then architecting an operating system with both its users and possible future uses in mind. This means Apple products have fewer security holes and are more "user friendly" than Windows.

And if it were an easy thing to do to plug all of the Windows security holes, I'm sure Microsoft would have done it by now. The problem comes down to the system "architecture", the foundational design of a software product.

Starting With Requirements . . .

This is not an exhaustive list but here are some basic requirements that we could start asking our elected representatives to put in place:

- Standardized frequencies between cars so that when they "talk" to each other they are literally on the same wavelength,
- Robust security designed into each and every computerized aspect of the vehicle so that it hacking opportunities are severely reduce – and a mechanism to alert the driver and provide traceability of the event if it is hacked,

- Clarity on whether it is acceptable to allow dealers and rental agencies to *ever* disable vehicles while being driven or even just parked,
- Clarity on who "owns" the data being collected by the black box that tracks driver activity and when it can and can not be used, and by whom,
- In the event of an accident, a clear, transparent process and technologies that help neutral technology experts and regulators understand the source of the problem to quickly determine driver or robot error,
- Specifications on how all software is to be built and confirmed as safe and accurate – including the software in a vehicle's "black box" and the mechanisms that interpret the data,
- Software updates occur only with the driver's explicit acknowledgement and approval (in a similar way to how computer software is updated only when the user inputs an administrative password accepting the approval),
- The ability for customers to download a list of all software updates made to their cars at any time and this information should be written in simple English,

Software and its supporting technologies are complex and require a joint industry-governmental-expert-and-*non-expert* effort to put together a meaningful suite of design requirements for the cars of tomorrow.

Some ideas for software standards can come from the FDA, for example. The FDA has software standards that the medical equipment industry must follow and this helps to *prevent* tragedies from happening. Problems still occur and the FDA tracks and monitors these problems. It can audit a company's processes as well.

When developing software for medical equipment, for example, if the company uses any other company's software, it must perform a full audit of the other company's software. Why? Because this informs their software engineers what they should do to ensure the safety of the code they're developing. Engineers need to build the new code to work properly with any "quirks" and unique features of the other company's software and to do that, a full audit of any third party software a company plans to use – or "talk" to – needs to be done.

As we know from the *Bookout v.Toyota* trial, despite the fact that Toyota used another company's operating system to support the execution of its own code, it did not audit the other company's operating system. Some of the software Toyota built contained flaws because it didn't examine its supplier's software.[166]

Unfortunately, there are no regulatory mandates that direct automobile companies to exercise the same care and caution when building the software in our cars as there are for companies building medical equipment.

Benefits of Clear, Directive
Software Safety Laws

Whenever clear, directive development rules are applied, the software development process is slower, more thorough and has more executive oversight and accountability. This creates a safer software product.

Without software safety laws, if one company embraces its duty as a corporate citizen, taking every possible step to reduce the risk of software defects in their vehicles, it will be at an economic disadvantage because there will be other car companies that just won't do that. "Faster, cheaper, better!" is the standard corporate mantra, but when it comes to the software in our cars, this mantra can result in "Faster, cheaper, deadly!"

And costly.

For sound automobile software, we need:

- A level playing field so that the "steady as she goes" development process applies to all automakers and no one can jump the gun and deliver less-than-sound software to the public, unfairly grabbing more market share while putting people's lives at risk,

- Enforceable *safety-first* software development standards, including the architecture, design, data capture and reporting tools that support security, safety and the ability of regulators to monitor automobile software compliance to the standards. This includes clear directives that specify thorough security, privacy and operating rules,

- Both Quality Assurance (checking) and Quality Control (testing) activities applied from "Ground Zero" onwards. Checks and tests are needed throughout the development process – not at the last minute after the developers have finished coding,

- Executive signoff confirming that the organization has followed the software development regulations, supported by a culture of quality from the executive floor to the shop floor,

- Chief Quality Officers reporting to the CEO who are *obligated by law to report any safety concerns jointly to the CEO and the regulator* – and a corporate culture that supports this level of due diligence,
- A genuinely independent third party certifying the new or changed software as having undergone all of the required safety standards during its design, development and testing,
- The ability for car owners to easily report their car's unusual behavior, both to the automaker and the regulator, so that patterns of software defects can be recognized and corrective action taken,
- A feedback loop to inform the regulators, automakers and the public about situations where software has failed so that consumers can protect themselves and the industry can continuously improve and strengthen its safety processes,
- A regulator fully empowered to monitor automobile safety and to take the necessary action when an automaker has failed to comply with the law,
- Those who run automobile companies held personally accountable when they fail to fulfill their public duty to protect customers.

In 2010, an electrical engineer who had experienced three separate incidents of SUA with his 2000 Lexus LS 400 said he was "convinced there is an electronic root cause" to the problem. He said to *CBS News*, "There needs to be designed traps to look for the condition – some sort of automated margin testing is needed to check whether there is a simultaneous braking while accelerating. The problem is that it will cost the industry millions, if not billions."[167]

Yes, it likely will. But now is the time to take charge of the future.

And if we do establish the proper software safety standards, we will have a transparent and ongoing quality improvement and communication process. As a result, this will help all automakers continue to strengthen the technology they expect us to trust with our lives.

Global companies can and do turn around the technology disasters they create, but it does take focus and it does take time. I have personally been involved in some successful turnarounds and know first-hand they require a tremendous amount of political will, sometimes starting at the board level. To make it happen with automobile software, this political will has to come not only from within the organization's leadership team but also from governments and regulatory bodies to encourage the people in those companies to do the right thing.

And to make that happen, those of us who drive or sit in cars need to care about our own safety, stay informed and put pressure on both the people producing our cars as well as our elected officials.

CHAPTER THIRTEEN

Creating Laws With Claws

__Fiduciary__ - the legal or ethical relationship of trust between two or more parties.[168]

If you or I knowingly sold products that were defective and killed people – and misled our customers into believing they were safe – we would get more than a finger wag and a tiny fine. We would go to jail. We'd lose just about all of our assets on fines and legal fees. If we'd been able to save our house through all the legal turmoil, we certainly wouldn't have a way to pay our mortgage because we'd be in jail. Our lives would never be the same.

Nor should they be.

We would have hurt not only the people who trusted us and died because of our products but also a huge circle of friends and loved ones. All of them – those who died, those injured, and those who must live with their terrible loss – would have deserved justice and justice is served when people responsible for causing harm to other people are held accountable.

The public does not truly know how much any of the top executives at GM and Toyota knew about their companies' decisions to sell cars with deadly defects. What we do know is that as executives they have a responsibility to ensure their companies follow the law. To help them pay attention to this fact, we need to hold them legally accountable to create and uphold a corporate culture that does not actively or even passively break the law. The media reports that GM, Toyota and even Honda have started to make the right sorts of changes, but how long will that last?

And what good will any of these changes be if so many companies are racing to build robot cars without any of us having a say in their foundational safety and security features? Even if every auto executive did his or her best to do the right thing, with so little legal oversight and so many loopholes, everyone can just throw up their hands and say "the technology was out of control and we couldn't help it."

We already know we have regulatory loopholes when it comes to software safety and, as Ryan Calo of the University of Washington has pointed out, we are heading straight into *many more loopholes* in the not too distant future, thanks to robotics. The combined challenges we face with artificial intelligence – machines making decisions – machines that interact with human life – means that we can't take the "driverless" car risks lightly.

Just because we can program a car to drive for you while you read a book or take a nap doesn't mean we should – until we have the proper legal framework in place to govern how a robot car is designed, built, tested and *certified safe* for people to use should we welcome these robot cars on our roads.

The problem is that, little by little, **robots are already driving our vehicles**. If a car is programmed to stop without you actually putting your foot on a brake, that's robotic action happening. But what if the car doesn't actually brake fully? Or stops when it shouldn't and you crash? Who is accountable?

And how can you find out what the robot did?

It seems to me that we need three types of "laws with claws":

- One set of laws to govern a corporation's fiduciary responsibilities to ensure that every possible action has been taken to keep people safe, holding individual executives accountable if the company breaks this important trust and sells products with deadly defects,

- One set of laws to govern mandatory software development standards with a clear suite of safety and security features that, once built, must be certified as safe, so that everyone plays by the same rules and there is a level playing field, and

- One set of laws to govern the "grey area" of robotics, defining who is accountable when software fails and how to build in reliable self-monitoring mechanisms to help determine where the "fault line" occurred between the various software components (and all the different companies that made the various bits and pieces of your car).

I am not a lawyer but it should not be only lawyers who influence how we govern technology. Citizens – people like you and me – need to be engaged so that we get the kinds of laws and governance that we and our children deserve. In this chapter are some ideas that we can start to work with that I believe can establish laws with claws – laws that have meaning and can't be wiggled out of so easily.

Existing Laws – So Far, So Bad

We are only now beginning to unravel the myriad ways that software flaws are impacting our lives and it will take time for the regulators to catch up. Even with today's more mechanically-focused regulations, we know that some auto companies routinely breach them. Although U.S laws demand that carmakers report known defects and deaths associated with their vehicles, they sometimes choose not to report – for *years*.

In 2014 alone, multi-national automobile companies in the U.S. admitted to breaking a number of regulations and were fined what amounted to less than a pinch on the nose. As mentioned earlier, the maximum fine that can be levied by the NHTSA is $35 million for any one breach. That year:

- GM was fined $35 million for failing to report known problems with their ignitions, problems GM had known about for ten years. *The $35 million fine represents less than 0.25% percent of GM's net profits for 2014.*
- Honda was fined $35 million for failing to report 1,729 deaths associated with its vehicles and another $35 million for withholding information about warranty claims. *The $70 million fine represents 4 days of earnings for the company.*[169]

While the U.S. Justice Department opened criminal investigations into both Toyota and GM, the number of Toyota and GM executives who were personally fined or went to jail?

Zero.

So, right now, if you're a carmaker, it certainly looks like it pays to break the law.

The behavior of automakers adopting software safety standards and following a safety-first architectural design won't change until the penalty is so severe that it can severely impact profits and impact the lives of the people making dangerous decisions that break the law.

U.S. President Obama would like the NHTSA's maximum fine to increase from $35 million to $300 million – still a laughable pittance for the automakers – but he is actually receiving pushback. *Reuters* reported that Republicans "have been unwilling to increase the maximum fine that NHTSA can impose".[170]

Party politics should not be playing with our lives like that. Other departments within the U.S. government, for example, are able to levy much more serious consequences against automakers than the NHTSA can with its $35 million ceiling. For example, in 2014, the Clean Air Act allowed the U.S. Department of Justice and the Environmental Protection Agency to make Hyundai and Kia pay "$300 million in federal penalties for 'systematically' and 'egregiously' overstating the fuel economy of nearly 1.2 million vehicles." [171]

So, break laws that harm the environment: $300 million. Break laws that kill people: $35 million.

I support protecting the environment. I also support protecting the lives of friends and family and people I don't even know.

Google Hires U.S. Regulator

Another comfy loophole that automakers – and now software companies – use is that they hire from the pool of staff at the NHTSA to then promote their own interests with the government. We saw that Toyota had former NHTSA staff acting as lobbyists managed to convince the regulators to halt four investigations into sudden unintended acceleration and their electronic throttles.

As I mentioned in an early chapter, **Google hired the Deputy Director of the NHTSA** – Ron Medford – as their new Director of Safety for Self-Driving Cars.[172]

There's a double whammy here:

Do you feel comfortable knowing that someone who had been in charge of a regulatory agency that didn't know how to monitor car software safety is now in charge of Google's self-driving car safety? Also, should there not be a "cooling off" period between the time when a regulator leaves his or her post and the time when industry can hire them to promote their corporate cause with the very regulatory agency they left?

Just as importantly, shouldn't someone who knows *how* to manage and monitor software safety be leading the charge at Google's Safety for Self-Driving Cars department?

Holding the Right People Accountable

Allowing business executives to make decisions that end up killing people without any personal consequences to themselves is more than a little bit unjust. It is criminal.

And as I've mentioned regarding how best to support quality management within organizations, without regulations leveling the playing field to ensure that all automakers are required to follow the same rules when building safety critical systems, executives who actually do care about protecting their customers are at an unfair disadvantage. It takes time and money to build safe systems. We have enough evidence to show that when it is easy for companies to take shortcuts and save money while not addressing defects that kill people, they do. And to date they have been financially rewarded for that.

We will get safer cars only when we:

- Make corporate executives personally responsible for their decisions, and
- Make it wholly unprofitable for auto companies to sell cars with deadly defects.

What if, for example, instead of being criminally fined $1.2 billion dollars for to hiding deadly safety issues with their "floor mat and sticky pedal" problems, Toyota had faced this?

- It was fined the *total profits* for the fiscal year 2014 (nearly $18 billion),
- None of the executives were permitted to collect that year's bonus,

- The company had to pay the legal costs of the regulator and/or Justice Department that had to pursue them in court (costs currently borne by the taxpayers), and

- All cases settled out of court – which Toyota nearly always ensures are kept confidential – are centralized with the regulator for its internal review and assessment when examining similar complaints in order to have the right information to conduct effective and proactive recalls. After all, patterns of court settlements tell a tale and regulators should always have the full story.

What if the specific executives – whoever they might have been – who were actively engaged in breaking U.S. law were sentenced to a minimum of one year in a *medium security* prison without a "get out of jail free" card? What if the people who decided not to fix things like GM's ignition switch problem were required by law to meet the children who had been sentenced to life in a wheelchair as a result of that "business" decision and can no longer play soccer or football or baseball or be on the swimming team?

In my opinion, executives would sit up and take notice and do their darned best to do the right thing. It would also cause shareholders and Board Members to actually start paying very close attention to safety issues to make sure that their company kept getting it right. Why?

Because it would certainly affect their bank accounts.

Right now, when an automaker of Toyota's stature breaks the law, even a $1 billion fine creates hardly a blip in their financials.

Software Safety Certification

Why should taxpayers pay to certify anything that a company produces as safe? Shouldn't the first step be to levy a cost on the company that inserts software into our cars?

Automakers should be the ones to pay for the certification process.

Of course, we know that the automakers will pass this cost along to the consumer but I, for one, would rather pay a little more money for my car to know that:

- The decisions its software will be making while I am in it will be good, sound decisions and
- Monitoring mechanisms are in place so that as the driver I am not automatically blamed for software failures.

Here are some basic things we need to do to help the automakers ensure they're building safer software:

- Require the NHTSA to act more like the Federal Aviation Administration (FAA). When airline accidents occur, the FAA collects the information and shares it across airline companies and manufacturers to help them improve their own operational safety practices. Doing this for software-related accidents would help the automakers make improvements, year by year.
- Ensure that there is an independent, effective software safety certification process in place whereby a third party, not the automaker or software company, confirms that the company building and/or changing the safety critical car software has followed all required regulations.

- Ensure that each automaker and any company building safety critical car software appoint a Chief Quality Officer who is accountable to report to the CEO as well as accountable to accurately report findings and safety analysis directly to the NHTSA.
- Ensure there are software safety standards and that the regulators have the ability to audit them.
- Ensure that any regulatory fine is more than a mosquito bite.
- Set out a strategy toward ensuring that automobiles have a secure technology architecture based on safety factors that all automotive companies must respect.

A sound architecture would build in manual overrides and safety features that could self-correct or intercept any error made by software. These are sometimes called "traps", built-in decisions that can take place when something starts to go wrong with the car. Some of this already occurs in our cars, but it has to be more robust and thorough – although it is costly to do.

And to avoid conflicts of interest:

- Ensure there is a "cooling off" period between the time a regulator leaves his or her post at the NHTSA and the time any company associated with safety critical car software can hire them.

Guiding Robotics Law

We've moved from the Industrial Age into the so-called Information Age, underpinned by an exciting explosion of technology. We are now racing into the Robotics Age and it's about time our lawmakers catch up. Ryan Calo, Associate Law Professor at the University of Washington, pointed out that unlike the Internet, robots can actually physically hurt us.

Robotics law is in its infancy and yet robots will soon reach into many more parts of our lives. We already have situations where, when software fails, companies are busy pointing at each other because the particular system that failed (such a banking system or public transit system or a "fill in the blank" system) was comprised of so many bits and pieces of other companies' products that no one can tell where the "fault" rests. Did it rest with the actual software code? Or did the "receiving" company not test it properly with their other unique set of systems?

And which company's executive(s) decided to cut safety corners to save money?

What if the software was designed to create its own unique set of decisions based on unforeseeable interactions with other systems?

These types of questions must be framed and answered – and answered proactively and consistently.

In his paper, The Case for a Federal Robotics Commission, Robert Calo sets out a framework and a rationale for urgently moving forward on setting up the proper regulatory framework for Robotics. As Mr. Calo points out in a paper published by the Brookings Institute:

"Technology has repeatedly played a meaningful part in the formation of new agencies."[173]   Listed below are those he mentioned in his paper:

- The advent of radio led to the need to manage the impact of radio on society and led to the U.S. Federal Radio

Commission, which "morphed" into the Federal Communication Commission (FCC). The FCC is now "charged with a variety of tasks related to communications networks and devices."

- The introduction of trains on the landscape "required massive changes to national infrastructure" and so the U.S. Federal Railroad Administration was established.
- When the small pox vaccine was introduced, society needed to organize a "massive outreach to Americans" which led to the United States Centers for Disease Control and Prevention.

And let's not forget that as more and more people began to fly on commercial airplanes, the U.S. government established the Federal Aviation Administration (FAA) in 1958. Over the years, it has been central to improving airplane safety and its very existence has both saved thousands of lives and supported the growth of a booming industry.

"Robots are organized to act upon the world physically, or at least directly. This turns out to have strong repercussions at law, and to pose unique challenges to law and to legal institutions that computers and the Internet did not," wrote Ryan Calo.[174]

A Robotics Law Center is required to help us manage accountability, but we don't need a separate one for automobiles and one for trains and another for medical equipment or refrigerators. The guiding principles of technology accountability shouldn't shift and change depending on a piece of equipment. Principles to guide technology's use and the related legal accountabilities should be based on democratic and judicial wisdom.

Right now, Google is partnering with major car manufacturers and expects to have self-driving vehicles on roads by 2020. That's not very far away. At this stage, with its financial might and influence, the company claims it "[doesn't] see any particular regulatory hurdles" and feels confident about putting cars with no steering wheels or brakes on roads that were not designed for cars driven by software.[175]

Why would they? After all, they hired someone from the NHTSA to be in charge of their driverless car safety program.[176]

Some U.S. states have pro-actively passed laws allowing driverless cars to be tested on roads. As highlighted in *The New York Times*, none of them have addressed legal issues such as: "What happens when a driverless car kills someone? Or less drastically, who pays the ticket when it doesn't notice a no-parking sign, or when an error in Google Maps sends it the wrong way down a one-way street?"[177]

We don't know what the future holds but we do know that robots will be there. It's time, now, to establish the legal framework that can also help both protect human life and support this evolving industry.

After all, we have safety standards for food quality, for construction and even for toy safety. Doesn't it make sense that our society adds software safety standards to that list?

CHAPTER FOURTEEN

The Courage to Care for Each Other

Software is becoming the invisible killer of the 21st century.

Without people, a corporation is just a pile of paper. Piles of paper don't make decisions that favor profit over people. Piles of paper don't issue painfully misleading press releases and lie to the public. Piles of paper don't actually care about money. Piles of paper don't have a responsibility to care for human beings.

People do.

It is the *people* within a company who collectively direct "the corporation" toward a good outcome for society or a bad one.

Whether we speak of the upper echelons of the most privileged and powerful automotive executives or the regulators or the shareholders of major automobile companies, all of them have families, drive cars, and care about the people closest to them. They cannot escape from defective automobile software any more than you or I can.

So it is all of us – you, me, our neighbors, friends, auto executives, politicians, and bureaucrats, everyone – who need to work together to reduce the risk of being killed by faulty software.

I encourage you to speak with your family and co-workers about this, send emails linking the web page at www.glitchwatch.com to help them understand what to do if they encounter SUA or software-related engine problems while driving. You can also print off the document and speak with your teenager about it, or "snail mail" the printout to friends and family. Just be sure that they know what to do – as best they can – if they are suddenly faced with a car that has decided to kill.

As I highlight in more detail in the appendices, I'd like to remind you that although cars with all the "shiny new software bells and whistles" look like fun, I'd recommend you do your best to find a car with the least amount of this sort of thing, and one with a manual transmission if you can . . . until cars are re-architected, redesigned, rebuilt and re-tested for safety.

In the meantime, reach out to your elected officials (including municipal officials who are encouraging companies to start testing robot cars on your city's roads) and blog about it and get your neighbors talking. These are our roads, our lives. Automakers and software companies do not have a fundamental right to expect us to sit in a car that can decide to kill. They do not have the right to put us at risk.

Whistles for the Whistleblowers

Until it becomes natural for a company to focus first on quality and public safety, the predictable turn of events when an employee tries to highlight a safety concern is what happened at GM: nothing at all is done and the person who had the courage to care is punished in some way. Although GM, Toyota and Honda all claim that they are changing their corporate cultures, what happens if they don't succeed? What happens if the same old "money above all else" mantra creeps back into their organizations?

We cannot let the automakers set out our car software safety framework all by themselves. Why not?

They're not good at it.

After all is said and done, for example, Toyota has an uneven whistleblower policy. In the UK, Toyota's whistleblower policy applies only to "theft or misuse of company property, financial mismanagement or corruption, or environmental issues"[178] – notice something missing in that list?

Public safety.

Over in India, however, although the Toyota whistleblower policy is also heavily focused on financially protecting the financial interests of company, it does have a little bullet that says its whistleblower policy also applies if an employee suspects "[h]arm to public health and safety".[179]

Society needs to protect the whistleblowers who go out of their way – at great personal cost – to do the right thing and warn the public about safety issues. Betsy Benjaminson's "reward" for undoubtedly saving lives and providing the U.S. Congress with documentation bringing much of Toyota's treachery to light is that not only is she being subpoenaed by Toyota, she now earns half of what she used to earn before she did the right thing.

Betsy Benjaminson, a single mother of four, refers to herself as "a gnat biting the elephant's toe." When Betsy broke her confidentiality agreement with Toyota[180], she said, "That was a difficult moment. Very difficult... I more or less betrayed [Toyota] in order to serve the general public."[181]

When I have discussed Betsy Benjaminson's courage with other people, I've had mixed responses. Most are proud of her humanity but there are others who feel she should be punished further because she – like all contractors and employees of large organizations – had signed a "non-disclosure agreement" and was legally bound to silence, though morally bound to speak.

Let me ask you this: if you were watching a group of people make decisions that you (and likely they) knew were putting others in mortal danger, would you remain silent?

Yes, I understand that Toyota has a right to subpoena or sue anyone who breaks a non-disclosure agreement but one must ask whether there are conditions under which individuals have a legal responsibility to do just that. In cases where someone highlights the fact that business decisions are being made that knowingly put people's lives and livelihoods at risk, doesn't it make sense that an honorable company would thank the whistleblower and even relieve her financial burdens rather than make her pay legal fees?

Surely Toyota's CEO, Mr. Toyoda, has the opportunity to do the right thing and thank Betsy Benjaminson for highlighting a corporate culture that curdled so badly it chose to break the law and sell deadly products, hiding this fact from its customers and regulators *for years*.

The U.S. Senate is now considering legislation to protect whistleblowers in the automotive industry. In fact, they are not only considering protecting them but also rewarding them, as the proposed legislation provides "employees and contractors for automakers, parts suppliers and car dealerships [to] receive up to 30 percent of penalties resulting from a federal enforcement action totaling more than $1 million – if they share original information on product defects or reporting violations with the U.S. Transportation Department or Justice Department." [182]

This is an excellent start and, hopefully, the U.S. government will find a way to help the whistleblowers like Betsy Benjaminson who helped Congress understand what Toyota and GM were doing.

Protecting whistleblowers does not negate the need for overarching software safety laws. The actual line of defective software code or design flaw is not easy to identify until it's too late – unless rigorous development processes are put in place to catch them early. A single individual on his or her own will seldom be in a position to pinpoint software safety issues without a defined and well-understood set of rules that everyone must follow.

If you would like to give Betsy Benjaminson a donation to help with her legal fees, please visit her website www.betsybenjaminson.blogspot.ca .

Our Collective Courage to Care

As I mentioned in the preface to this book, we are all equally vulnerable to the effects of faulty automobile software but we are not helpless. We can each do something about it right now and bit-by-bit we can lay the foundation for safer software in our cars.

We must all encourage our governments to make laws, then, that *hold both the corporations and the people who make deadly decisions accountable*, and properly protect the whistleblowers. Again, it is not piles of paper that are accountable to exercise their fiduciary responsibilities, but people.

And when people, like Betsy Benjaminson, have the courage to care, they absolutely should not be punished for it.

We live in a democracy. A democracy is supposed to be a place run "by the people, for the people." "We the people" have the right to demand that governments govern and hold those who make decisions that harm others accountable, even criminally responsible.

This democratic world of ours didn't pop out of some mythic womb. It has been shaped and guided by people who lived in our past and cared about the future, people who worked hard to give us the right to vote, the right to demand that those we elect attend to our collective well-being, and the right to expect that products sold to us don't kill us.

The Road Ahead

It is true that in this increasingly complex world, we do not yet have the ability to build perfect software. But we can certainly create *safer* software.

When developing a strategy to improve software quality, it's important to acknowledge that it takes time and a highly focused effort. It's also important to remember that there are things that can be done in the short term and those that will take years and possibly decades. When helping organizations improve how they deliver quality to their customers, I like to look at the "quick wins" – what can be done now – as well as the longer-term roadmap to better, more robust software.

Everyone who drives or sits in a car – or simply cares about someone who does – should call upon our governments and insist that we now take those steps, both the short-term "quick wins" and set out the longer-term strategy. We need to ask automakers to actually demonstrate the principles of caring for their customers, principles that they so easily claim in their press releases and have too often shamefully ignored.

In a democracy, power is never seized – it is ceded by its citizens. As we step into the The Age of the Robot, we owe our children and future generations laws that protect them from dangerous cars.

The future will be driven by software.

Together, let's make it safe.

APPENDIX A - PROTECTIVE MEASURES

When it comes to software safety, we need to do more than pray. We need to take action and make sure that governments around the world institute *laws with claws*. These software safety laws would regulate how software is built and invoke serious financial penalties for those companies caught breaking the law.

But building better laws and regulating software safety can't be done overnight. We need to *make every possible effort right now* to protect ourselves as best we can.

Those who initially reviewed this book before publication were not happy to know what I had learned about the software in our cars. "You've got to tell us what we can do to protect ourselves," they said, insisting I provide a clear set of steps that all of us can take if we encounter software problems while driving – and they were right. The truth is, though, I cannot provide the breadth of protective actions because there are just too many variations of possible things that could go wrong.

But to address the two most deadly software-related car problems, I searched through and summarized various automotive blogs and listened to a 911 call – with a happy ending – where the operator was able to successfully talk a terrified mother through the steps she needed to take to stop her car as it raced out of control through one red light after another, with her baby in the back seat.

Probably the two most urgent things for you to know are how to wrestle back control of your vehicle if it:

- Suddenly speeds up and you can't make it stop. We now know this problem happens in more than just Toyotas,
- Your engine suddenly shuts down as you're driving.

Both of these situations can be deadly, especially if you panic. So the first thing to remember Rule #1 in *The Hitchhiker's Guide to the Galaxy*: **Don't Panic**.

I have also listed this information on www.Glitchwatch.com so that you can send links to friends and family or just print it off and have a conversation with your teenager or neighbor.

At Least KNOW This: How to Save Yourself

Car Suddenly Accelerates

SUA – Sudden Unintended Acceleration. Occurs when the vehicle suddenly speeds up. When this happens, typically the brakes don't work, not even the emergency brake. Cars can reach speeds of over 120 mph. (190 kph.).

If you experience SUA, remember this: *if you can't get control of the vehicle, put the car in neutral and see if that helps. If it doesn't, immediately turn the engine off.*

1) Remain calm and stay focused. Be aware of traffic conditions around you.

2) Do not pump the brakes. Press them once and if traffic conditions permit, press them *hard* but do not pump them – pumping them seems to rob them of strength.

3) Put the car in neutral and turn the ignition **off**

   a) If you have a "push button" ignition, press the push button ignition for as long as it takes to turn the engine off.

4) Put on your flashers and then slowly *move to the shoulder* of the road (if you can do so safely) or stay put and wait for help as you sit in your car in the middle of the road.

5) Call for a tow truck (or 911 if someone is hurt or there has been a collision) and

6) Report the problem to the correct government authority and provide a copy to your lawyer (or trusted family member) in case you don't survive the next time your car "decides to kill".

Now, remember, with your car turned off, a lot of things won't work, like airbags and other safety features (even though *The Wall Street Journal* reports that airbags are supposed to work even after power is lost.[183]) But airbags didn't protect Mark Saylor and his family, nor did they protect the many others who died because their cars decided to kill.

Cars Rolling Away or Engines Turning Off

In some Hyundai, Ford[184], GM, Chrysler and likely other manufacturers' vehicles there have been mechanical and/or software problems either with their ignitions or transmissions that:

- Cut power to the engine while you are driving, which then causes the steering and brakes to malfunction and the airbags not to deploy.
- Allow cars to just roll away while parked (hence, the value of always, always putting on your parking brake).

Whenever you park your car

Absolutely every time you park your car, ***put your parking brake on*** *(special note: some parking brake are also controlled by software, so it is not a failsafe method to protect your car from rolling away while turned off and parked).*

If your engine suddenly stops while driving

Having your engine stop and your brakes not work very well if you're going up- or downhill can be very traumatic. If this happens:

1. Remain calm and focused.
2. Remember that your brakes will no longer have safety features like ABS and traction control so be cautious about how you engage the brakes. Try not to slam on the brakes unless you feel you really must.
3. If you need to press the brakes, press them and *keep them pressed* – pumping them seems to rob them of strength.

4. Put your flashers on and try to pull over to the side of the road, if you can.

5. Call for a tow truck (or 911 if someone is hurt or there has been a collision) and

6. Report the problem to the correct government authority and provide a copy to your lawyer (or trusted family member) in case you don't survive the next time your car shuts itself off while you're driving in traffic.

At Least DO This: Protect Yourself

Reduce the Risk of Deadly Car Software

Here is a list of things that you can do to reduce the risk of owning a vehicle with deadly software. I've also listed (with some effort, as it wasn't straight-forward) where you can go to report any unusual behavior.

Reporting unusual car behavior:

- If you're reading this book on an e-reader, you can click on the appropriate links at the Glitch Watch website.
- If you're reading this book in paperback, I recommend you go to your computer and find the page at www.glitchwatch.com where there are links to various countries' federal agencies to report a problem with your car.

When I searched for places to report problems, I found a number of sites that seemed to be controlled by the auto industry.

While it's fine to report to the auto industry, *first* report suspicious behavior to your country's federal government authorities.

We do need a much better way to consolidate and study the information about software problems related to our cars. Until then, we need to use the reporting mechanisms that we have in various countries and we need to press for change.

#1 - Say "no" to software bells and whistles.

Before buying a new car, check at the U.S. Government website SaferCar.gov and review the car manufacturer's history of recalls and complaints, especially with respect to the model you're thinking of buying.

When buying a new vehicle, select the one with the least number of software "bells and whistles" and be sure to tell the sales person and the dealership why you want less software, not more.

If at all possible, try to get a car with a manual transmission. You will have more control over the vehicle with a manual transmission because you are able to slow down by "down-gearing" rather than pressing the brakes.

One of the first facts to understand with software is that the more complicated it is, the more likely it will break (unless it's undergone *extensive* testing).

Think of a chain. If you have two links, there is only one place where the chain might break. If you have 40 links, there are 39 places where the chain could break. Today's software is a complex web of interactions – from the code, itself, to radio frequencies that trigger unexpected behavior.

So, **Apply the KISSS Principle** (Keep the Software Simple, Stupid) when buying a car and you might be saving your life.

#2 - Understand what the heck the software you're stuck with is actually doing.

Even cars that aren't fully loaded will have a lot of bits and pieces that are run by software. Remember, software is basically a decision-making tool. It's designed to make decisions for you. In Jean Bookout's case, her car's software decided to take control of the vehicle and there was nothing she could do about it.

Take the time to understand what the software in your car is doing and where things might go wrong. Ask the dealership to help you understand all of the software that controls your vehicle.

These days, the list of software control points in any given vehicle will include a lot more than you would imagine.

#3 - Report all unusual vehicle behavior.

If the car "mysteriously" does something (like speeds up without you knowing why, or turns itself off, or the brakes temporarily fail, or the doors pop open while you're driving), make a note of it and immediately report it to both the dealer and to car safety agencies, both government and citizen-based. You can find a list of the various agencies on the Glitch Watch site.

Keep a record of this unusual behavior and the fact that you did, indeed report it.

Why?

Three reasons:

1. Automobile companies are required by law in the U.S. and other jurisdictions to report fatalities and known

problems associated with the operation of their vehicles (and they sometimes break this law).

2. The more information gathered on different vehicles and different types of software problems, the easier it will be for car manufacturers to fix the situation in the long run. It helps them perform "root cause analysis" which then helps them improve both the software as well as the software development and governance processes.

3. You may not survive another software failure and your family needs to have a record of your car's unusual behavior.

Some government agency sites ask you to enter your Vehicle Identification Number (VIN). In this case, your lawyer and/or family can quickly find what you reported.

#4 - Advocate for software safety laws.

Support citizen action groups calling for software safety laws and contact politicians to explain why it's important to take action now. Over time, I will post links to citizen groups that are working hard for safer cars at the Glitch Watch site. For now, here are a few ideas that people could begin to encourage our legislators to make happen:

a) All vehicles should come with a simple description of each mechanism that is controlled by software and what it does (and what it shouldn't do).

b) All auto manufacturers should be required by law to build software according to a common architecture and set of software development standards that are federally regulated and then go through a safety certification process. This should not be a "voluntary" or optional process.

c)   Companies found to have breached item b), above, should be fined. If their bad software has killed, then they should be fined not a measly percentage of that year's profits, but a significant amount of money that seriously eats into their profits. It should simply not be profitable to kill and, yet, in the auto industry, it is.

d)   Regulators should provide a simple way for citizens to report suspected and known software problems in order to build a body of evidence and keep proper records. This information should be made available to support lawsuits and to help automobile companies build a better understanding of the "root cause" of their software problems.

e)   Do not allow automotive regulators to work for or advocate on behalf of automotive companies for five years after they have completed their time as government regulators.

f)   Similarly, ensure that there is a considerable amount of time before industry representatives – retired or otherwise – get appointed to any regulatory body.

#5 - Forewarn friends and family.

Tell your friends and family and co-workers and people at the grocery store – anyone you come across who drives a car or sometimes sits in one (in other words, everybody!) – about the fact that broken software in our cars is now killing people and that we have to do something about it.

And ask them to take the steps necessary to try to protect themselves and those they love by taking the preceding steps, too.

I have provided a document listing the steps people can take to protect themselves on the Glitch Watch site. Please feel welcome to download it and print off or send it to people you know.

In my opinion, Driver Education courses should now include information about what to do when the software in your car behaves erratically, putting lives at risk.

APPENDIX B –
MORE INFORMATION

This set of appendices has been provided so that you can:

- Read through some comments by embedded electronics expert, Michael Barr, on his assessment of the NASA report regarding Toyota software.

- See samples of complaints filed with the NHTSA against both Toyota and other manufacturers regarding their SUA experiences.

- Read more details on Toyota's twists and turns regarding SUA

- Meet a few whistleblowers that advocated – at personal loss – for our safety with respect to Toyota and GM cars.

- Read the curious details about how Mazda solved a problem of real bugs clogging up their engines (because sac spiders evidently like gasoline) using an ingenious software solution.

Barr's Review of NASA Report

As I've said before, NASA was unable to find the SUA coding problem that The Barr Group found and *admitted that just because they couldn't find it, it didn't mean it wasn't there*. Yet the media, to this day, will only highlight the first bit: that NASA couldn't find the problem.

Why couldn't they find what Michael Barr found?

For one, their scope was limited. Two, they had a limited amount of time.

Before he had concluded his own examination of Toyota's embedded electronics code for the landmark 2012 economic loss class action against Toyota and then the 750-page report for the prosecution in *Bookout v. Toyota* in 2013, Michael Barr analyzed the NASA report and, in his Embedded Gurus blog dated March 1, 2011, said, "It looks to me like Figure 6.2.3-1 of the NASA Report (p. 30) shows that UA complaints filed with NHTSA increased in the year of introduction of electronic throttle control for the vast majority of Toyota, Scion, and Lexus models . . ."

However, he also noted two key observations about the NASA report – aside from the fact that there were a number of redacted statements that made it difficult to fully comment on the whole report. I have left the links in his text that guide you to his further, more technical, explanations as well as quoted him virtually word-for-word.

Observation 1: Stack Flow Analysis

Michael Barr noted that the NASA report stated that:

- "Toyota designed the software with a high margin of safety with respect to deadlines and timeliness. ... [but] documented no formal verification that all tasks actually meet this deadline requirement" and
- "All verification of timely behavior is accomplished with CPU load measurements and other measurement-based techniques."

"It's not clear to me," continued Barr, "if the NASA team is saying it buys those Toyota explanations or merely wanted to write them down. However, I do not see a sufficient explanation in [the NASA report stating on page 132 that] "The [worst case execution time] analysis and recursion analysis involve two distinctly different problems, but they have one thing in common: Both of their failure modes would result in a CPU reset. ... These potential malfunctions, and many others such as concurrency deadlocks and CPU starvation, would **eventually** manifest as a spontaneous system reset." (Emphasis added)

Michael Barr's response to page 132 of the report: "Might not a deadlock, starvation, priority inversion, or infinite recursion be capable of producing a bit of "bad behavior" (perhaps even unintended acceleration) before that "eventual" reset? Or might not a stack overflow just corrupt one or a few important variables a little bit and that result in bad behavior rather than or before a result? These kinds of possibilities, even at very low probabilities, are important to consider in light of NASA's calculation that the U.S.-owned Camry 2002-2007 fleet alone is running this software a cumulative one billion hours per year."

Observation 2: Other Factors

Upon reflecting "on the steps NASA might have taken in its review of Toyota's ETCS-i firmware, but apparently did not," Michael Barr wrote a blog in March 2011 title, "Unintended acceleration and other embedded software bugs". In it, he pointed out that "there is no mention anywhere (unless it was entirely redacted)" of the following:

- rate monotonic analysis, which is a technique that Toyota could have used to validate the critical set of tasks with deadlines and higher priority ISRs (and that NASA could have applied in its review),
- cyclomatic complexity, which NASA might have used as an additional winnowing tool to focus its limited time on particularly complex and hard to test routines,
- hazard analysis and mitigation, as those terms are defined by FDA guidelines regarding software[185] contained in medical devices, nor
- any discussion or review of Toyota's specific software testing regimen and bug tracking system.

"Importantly," Barr continues, "there is also a complete absence of discussion of how Toyota's ETCS-i firmware versions evolved over time. Which makes and models (and model years) had which versions of that firmware? (Presumably there were also hardware changes worthy of note.) Were updates or patches ever made to cars once they were sold, say while at the dealer during official recalls or other types of service?"[186]

All of this feeds into a full and complete understanding of the source of electronics problems.

Sampling of SUA Complaints
Filed with NHTSA

I have provided just a sample of the 1,841 complaints related to cars that suddenly accelerate (February 2015). Of the 1,841 complaints currently filed with the U.S. NHTSA, 41% of them are related to Toyota and the rest apply to a long list of carmakers.[187]

Toyota SUA Complaints

NHTSA Complaint #10663576 - Toyota

> *In June 6, 2011 I had a crash because my Toyota Corolla suddenly accelerated as I was turning into a parking spot. Toyota had the car inspected in house and claimed there was no defect in the car's systems. I filed a report with the NHTSA to no avail. Just recently on December 13th & December 14, 2014 the same thing happened. The car suddenly accelerated as I was turning into parking spots. I am positive there is a defect in the car.* **I am afraid I may be killed or injured** *with this sudden unintended acceleration.*

NHTSA Complaint # 10550517 - Toyota

> *When slowly pulling into a parking space in front of a building, the* **vehicle suddenly accelerated across a 3-foot wide sidewalk and into one of the storefronts of a building** *– resulting in extensive damage to the front of the vehicle (currently in the body shop) and extensive damage to the storefront. Fortunately, no one was on the sidewalk at the*

time. This is the first sudden unintended acceleration for us . . . but apparently not for the 2009 Avalon.

NHTSA Complaint # 10543880 - Toyota

*On the evening of August 30, 2013, my wife and I were involved in an auto accident due to a Sudden Unintended Acceleration problem with our 2002 Toyota Camry. I was driving at a low speed around a cul-de-sac when **the car accelerated out of control and crashed into a tree.** I pressed the brake a few times but it was ineffective. The crash destroyed the car and is determined to be a "total loss."*

NHTSA Complaint # 10537281 – Toyota

*I experienced a sudden unintended acceleration on my 2013 Camry SE on 08/20/2013. **The car hit a big tree and was not repairable.** Totalled. Thanks to the airbags, I only had a couple of bruises here and there.*

SUA Complaints – Other Automakers

NHTSA Complaint # 10462921 - Chrysler

While driving on U.S. Highway 270 in Arkansas this date, my 2009 Jeep Grand Cherokee experienced sudden unintended acceleration. I was pressing slightly on the accelerator pedal while going up a hill when the engine went into a wide open throttle condition. I was able to maintain control of the vehicle and turned off the ignition, allowing the vehicle to coast down to approximately 45 mph. Upon turning the ignition back to "on" the vehicle again experienced SUA and accelerated quickly to 60+ mph. I again turned the ignition switch to the "off" position and allowed the vehicle to coast back down. Upon turning the engine back to "on", the engine went back into a wide open throttle condition again. After four tries (turning ignition "off", then "on", the engine rpm returned to a normal operating condition and I was able to continue with my trip. There was no floor mat interference with the accelerator pedal. The speed control system was not activated. I am a pilot with 500 hours of flight time and recent annual check ride qualification/training to handle emergency situations. Had my wife or the average driver been operating the car, there would likely have been a serious accident. **The vehicle simply went berserk.**

NHTSA Complaint # 10577126 - Honda

Two incidents of sudden unintended acceleration. The first on January 31, 2014. The second on March 15, 2014. Both incidents occurred at a very low rate of speed of approximately 5 mph. In both situations the car was coming to a stop with the brake being applied. Suddenly, the RPMs increased and the car moved forward with the brake unable to stop the car even though both feet were applied to the brake pedal. In the first incident, the car was stopped by putting the gear shift into neutral and turning the engine off. In the second instance, the car was pulling into a parking space when **the engine revved up, jumped a curb and hit a small tree**, *as soon as the tree was struck, the engine returned to normal speed. This has been reported to Honda of America as well as to the local Honda dealer. The car was brought to the dealer two times but the dealer was unable to diagnose the problem. Honda of America has opened a case but thus far has taken no action. This is potentially a very dangerous situation.*

NHTSA Complaint # 10662527 - Hyundai

We own a 2011 Hyundai Santa Fe and had an incident last night pulling into a parking spot, which appears to be something others have experienced: SUA Sudden Unintended Acceleration. **I was pulling into the parking space with my foot on the brake and the motor just started racing full throttle for no apparent reason.** *I mashed on the brake and felt it fade out under the force of increased motor RPM. I quickly shifted into neutral and the engine continued to race but I had my foot completely removed from the brake or pedals. I then shut the engine off. My wife and I were shaken but uninjured. After completing our reason for the stop, the vehicle started normally and we returned home with no further incident. But our confidence in the vehicle has gone to zero.*

NHTSA Complaint # 10587944 - Mercedes

My 2013 Mercedes E350 W4 exhibited very high engine revving followed by high acceleration which could only be controlled by activating the parking brake. This is the second dangerous incident that has occurred with this car.

NHTSA Complaint # 10587944 - Mercedes

The contact owns a 2013 Mercedes Benz E350. The contact stated that the vehicle experienced unintended acceleration while stopped at a traffic stop light. The vehicle would not immediately respond after depressing the accelerator pedal but would suddenly accelerate forward. Additionally, the contact indicated that the failure sporadically occurred when the vehicle was in reverse. The backup camera had also become inoperable. The vehicle was taken to the dealer several times but the failure could not be diagnosed. The manufacturer was notified of the failure. The failure mileage was 3,000. Updated 08/18/14 – The consumer stated the dealer drove and inspected the car. The acceleration was felt but did not program to the vehicle's computer. Updated 9/5/2014.

NHTSA Complaint # 10629613 - Volkswagen

*On Aug. 09 2014 coming to a stop I applied the brakes. All of a sudden the engine started to accelerate uncontrollably trying to move the car forward. I pressed the brakes as hard as I could and I stopped the car after I hit the car in front when I was able to turn the ignition key off. This sudden unintended acceleration malfunction also happened on Jan. 29 2006 and Jan 30 2011, also reported to NHTSA at those times. **It is a life safety issue and should be addressed A.S.A.P.** The manufacturer refuses to take any action.*

NHTSA Complaint # 10588115 - Ford

*Vehicle had a sudden unintended acceleration event which nearly resulted in an accident. This was a very nerve-racking experience and I had difficulty stopping the vehicle (despite applying full brakes) as the RPM continued to increase to approx.. 5000. On stopping the vehicle I managed to place it in neutral and the engine continued to rev excessively. I turned off the engine. Vehicle was brought to local ford dealership who confirmed **a high rev problem with the electronic throttle control** and recommended a "throttle body" replacement.*

Toyota's Twists and Turns – Detail

The detail, below, provides a little more background to the chapter, *Toyota's Treachery*, and contains references to the source material where you can look for additional information. To keep the timeline intact, in some cases it repeats the information in that chapter.

- 2001 – The 2002 Toyota Camry is substantially redesigned. According to Safety Research & Strategies Inc., new or revised vehicle systems included:
  - The electronic throttle control system (ETCS-i) and
  - Transmission and braking systems which consisted of "an accelerator pedal sensor, a throttle control motor, a throttle position sensor and the engine control module (ECM)."[188]

- In 2002, a Toyota Camry owner files the first complaint about sudden unintended acceleration with the NHTSA. Toyota issues Technical Service Bulletin TSB EG017-02 to update the Electronic Control Module calibration to address 'engine surging' on 2002 Camrys . . . The Engine Control Module (ECM) calibration [was] revised to correct this condition." [189]

- Since 2003, the U.S. NHTSA has been involved in eight investigations involving Toyotas and SUAs, seven investigations on Toyotas equipped with Eletronic Throttle Control (ETC) systems.[190]

- Six of the eight investigations were closed and two resulted in recalls involving floor mats (2008) and pedal entrapment (2009).[191]

- Government and legal documents indicate that "[a]t least four U.S. investigations into unintended acceleration by Toyota Motor Corp. vehicles were ended with the help of former regulators hired by the automaker, warding off possible recalls, court and government records show . . . *All four of the probes the Toyota aides helped end were into complaints that the unintended acceleration was caused by flaws in the vehicles' electronic throttle systems.*"[192] [Italics added.]

- In 2008, top engineers and management at Toyota in Japan were very concerned about their Imperial Family's "speed control" problems with their luxury Toyota. There were concerns about the heir to the throne's safety, and something had to be done.

  "The problem seemed rooted in electronics — but its solution was elusive, even to all those trained minds. [For the Imperial Family] Toyota **replaced the gas pedal, the throttle system and the engine computer** at its own expense." [193] [Emphasis added.]

- In 2010, *Bloomberg* reported that Toyota spokesperson, Martha Voss claimed that Toyota's lobbying activities to stop investigations "have been consistent with our efforts to maintain the highest professional and ethical standards in all of our legal and regulatory practices. Their paramount concern was for the safety of every single owner of one of our vehicles."[194]

Remember, that was 2010 – read below to understand what Toyota's "highest professional and ethical standards" turned out to be between 2011 and 2014.

- By 2007, Toyota knew that their floor mats had problems but kept this to themselves and continued to sell vehicles and claim driver error. A little while later, they understood that they needed to urgently make changes to the design of their accelerator pedals and began the process of making these engineering changes in Europe but later decided against applying the fix in North America.[195]

During this period, a man in Minnesota driving a Toyota was wrongfully convicted and sentenced to eight years in jail as a result of what he claimed to be his vehicle's brakes failing. (He had crashed into the back of another vehicle, killing three occupants.) Mr. Koua Fong Lee served two years before being released when Toyota began recalling their cars due to floor mats and sticky pedals.[196]

- Through most of 2009, Toyota continues to claim that driver error is responsible for the sudden unintended acceleration problems *until* the horrific accident involving Mark Saylor and his family in August of 2009. Listening to the 911 call that was made from the doomed vehicle during the last minute before the crash, which killed Mark and his three passengers, including his 13-year-old daughter, it became clear that this was certainly something more than driver error. In *November 2009*, after the well-publicized tragedy, Toyota begins to recall some of its vehicles and attempts to address SUA by replacing floor mats.

- In January 2010, Toyota issues a second recall because it had become clear to the public that the floor mat replacement did not solve Toyota's SUA woes. This time, the recall addresses the "sticky pedal" problem (note that Toyota had identified the sticky pedal design issue at least one or two years earlier, but chose not to fix or report the problem, as they are legally required to do).[197]

- The *Los Angeles Times* reports that U.S. Attorney Preet Bharara said, "In the midst of the firestorm over the San Diego accident [the one involving Mark Saylor and his family] and unwanted acceleration problems, what did Toyota do? It quietly canceled the 'sticky pedal' fix in America . . . Toyota executives directed employees not to put anything about the cancellation in writing and to avoid a paper trail."[198]

- In 2010, the NHTSA reports that they had received over 6,200 complaints regarding SUA in Toyota vehicles for period 2000 to mid-2010. During that time, they attributed 89 deaths and 57 injuries to SUAs in Toyota vehicles.[199]

- In 2010, *Bloomberg Businessweek* reports that former NHTSA regulators hired by Toyota helped to end four of the eight NHTSA investigations into the company's SUA issues. All four of the "halted" investigations were related to problems with Toyota's electronic throttle systems.[200]

- In August of 2010, referring to a report by the NHTSA, the U.S. Transportation Secretary Ray LaHood pronounces, "The verdict is in. There is no electronic-based cause for unintended high- speed acceleration in Toyotas. Period."[201]

- In two separate instances in 2010, NHTSA fines Toyota $48.8 million for failing to report known problems in a timely manner.[202]

- In 2010, Toyota announces that it had a software bug in the laptop computer software that it has been using to interpret the results of the black box device from accidents involving their cars.[203]

- In 2011, NASA reports that it could not find the software code that is the cause of the SUAs but did point out (although it is seldom reported this way) that just "[b]ecause proof that [Toyota's] ETCS-1 caused the reported UAs was not found *does not mean it could not occur*."[204] [Italics added.]

*** Right at this point (February 2011), *Bloomberg*, *Businessweek* and *NBC* publish a piece by Ed Wallace, a car reviewer for *Fox*, titled: "Toyota, The Media Owes You an Apology". He claimed that because NASA couldn't find a problem with Toyota's software, there was no problem and that Toyota's reputation had been tarnished for no good reason.[205] [Unless, of course, you note the bits about lying and breaking the law that they finally admitted to . . . – author's note.]

A few days later, Canada's national newspaper, *The Globe and Mail* restated much of Mr. Wallace's uninformed commentary.[206] Neither *The Globe and Mail* nor Ed Wallace went to the effort to point out that NASA had made it very clear that just because they couldn't find a problem didn't mean that one didn't exist in Toyota's software.

It seems that NASA had a time-limited mandate with a scope that didn't allow them to find what embedded electronics expert, Michael Barr, was then able to find. ***

• In 2012, Toyota settles a class action lawsuit in excess of $1.6 billion[207] for "loss of value on vehicles affected by multiple recalls and [to] install special safety features."[208] According to the *EE Times*, testimony from the embedded electronics firm, Barr Group led to this "economic-loss settlement by Toyota . . . Because of that settlement, details of the technical discoveries made back then by the experts were not made public until the Oklahoma trial [in 2013]."[209]

The settlement does not cover wrongful death or injury and Toyota continues to work through hundreds of lawsuits. This $1.6 billion settlement also does not cover lawsuits that claim faulty software or electronics.

The *EE Times* also reports "[s]imilar testimony [to that of Michael Barr] and extensive software analysis reports had been filed previously in other courts looking into unintended acceleration. But none of that material became public, because *Toyota paid settlements and obtained gag orders before those cases went to trial*."[210]

• Toyota wins many unintended acceleration cases by blaming the problem on driver error and pointing to the fact that NASA could not find the software problem but in cases where "the plaintiffs' attorneys submitted a full report on a flaw in the vehicle's electronic throttle control system, a pattern emerged: Toyota opted to settle before the case went to trial."[211]

• In 2012, **The New York Times** reports that after a long investigation, "government officials concluded last year that there was no evidence that faulty electronics systems contributed to the acceleration issues. *But a subsequent*

156

review of that inquiry by a branch of the National Academy
of Sciences found that federal regulators had lacked the
expertise to monitor electronic controls in automobiles."[212]
[Italics added.]

- In 2012, NHTSA again fines Toyota for failing to report
  known problems in a timely manner. This time the fine is
  $17.4 million.[213]

In March 2013, while driving to work, Massarat
Chaudhary's 2009 Camry suddenly accelerates. She calls
her daughter and explains what happened and that her
brakes don't work. The vehicle crashes into the Sacramento
River in California. While trapped in her vehicle with the
water rising, Mrs. Chaudhary is able to dial 911 to call for
help, explaining that she can't get out of the car. The
operator can hear her pounding on the window, trying to get
out, but she is unable to escape and after a few minutes, the
line is dead. Mrs. Chaudhary drowns in her vehicle before
help can arrive.[214]

- In October 2013, in support of the lawsuit in Oklahoma in
  the *Bookout v. Toyota* case, embedded electronics engineer,
  Michael Barr, and his team "uncovered gaps and defects in
  the throttle fail safes" of Toyota software and were able to
  replicate the SUA using a Toyota vehicle . . . Stack
  overflow and software bugs led to memory corruption. .
  And it turns out that the crux of the issue was these
  memory corruptions, which acted [according to Barr] "like
  ricocheting bullets."

In addition, Barr reported that they had also examined the source code for the black box and "found that it can record false information about the driver's actions in the final seconds before a crash."[215]   (Remember that in 2010, Toyota admitted that it found a bug in the laptop computer that read crash data from the black box...which, we learn in 2013, also had a bug.)

Toyota lost the case and then settled with the victims' families, yet it continues to deny that SUAs can be triggered by faulty software in their electronic controls. According to the terms of the settlement, The Barr Group's report cannot be made public.

- Soon after the verdict in favor of Jean Bookout, Toyota entered into an "intensive settlement process" with "the hundreds of U.S. state and federal lawsuits alleging that defects caused its vehicles to accelerate suddenly and crash, resulting in serious injuries and deaths."[216]

- *The Economist* reported in 2014 that "[o]ne internal [Toyota] memo . . . showed the company crowing about having talked the National Highway Traffic Safety Administration out of ordering a recall, saving Toyota millions of dollars."[217]

- In 2014, in "a landmark settlement of *criminal* charges, Toyota Motor Corp. admitted deceiving regulators about deadly safety defects and agreed to pay $1.2 billion, the largest penalty ever imposed on an automaker. In the unprecedented deal with the U.S. Justice Department, the world's largest automaker admitted it misled consumers about two defects that caused unintended sudden-acceleration incidents — sticking gas pedals and floor mats trapping the pedals."[218] [Italics added]

- Of special note, prosecutors forced Toyota to agree not to take the $1.2 billion fine as a tax-deductible offence, which saved the U.S. taxpayers approximately $420 million.

- The settlement is related only to the floor mat and sticky pedal flaws, not software.

- Reuters noted that "[a]lthough no individuals at Toyota were charged, the case was the first federal criminal case of its kind since the passage of the first U.S. auto safety law 48 years ago." [219]

- In 2014, Toyota Corolla owner, Mr. Rugini, filed a complaint because his car had had 3 separate SUAs *after* having the floor mats and sticky pedal fixed, with the third SUA resulting in a crash. An electronics expert himself, Mr. Rugini reached out to Sean Kane, an automobile safety expert. Junko Yoshida of the *EE Times* reported that Kane indicated he is aware of 164 complaints that are very similar to Rugini's – floor mat and sticky pedals have been fixed, yet the vehicle still suddenly accelerates on its own.[220]

- To date, Toyota does not admit that its electronic throttle has any software problems that can cause SUA.

Some Whistleblowers

At both Toyota and GM, people who had the courage to care about more than just their paycheck came forward to warn the public about serious problems within those organizations that put people's lives at risk. None of them have been rewarded for their care and concern. In fact, the two Toyota examples I list here have been sued by Toyota.

Betsy Benjaminson

A single mother of four, Betsy Benjaminson is a Japanese-to-English translator who was working for a New York law firm representing Toyota. She had access to hundreds of documents and as she read through them, she became concerned that what she was reading in the engineers' notes to management was not matching Toyota's PR spin on the SUA situation.

When she learned of yet another horrific SUA situation (where a Camry suddenly accelerated and lunged over a cliff), she knew that she had to do something, so she spoke to her Rabbi to discuss her situation and then met with a lawyer, understanding that she was putting her livelihood at risk.

She then provided documents to the media that showed how Toyota spent a considerable amount of time investigating and then changing the Japanese Imperial Family's luxury Toyota software while claiming there were no software problems in other vehicles. She came forward publicly and appeared before the U.S. Congress.

With this information, the U.S. Congress was able to see how Toyota's management had been misleading the public and the U.S. regulators about their deadly vehicles.

Toyota is suing her.

Dimitrios Biller – Former lawyer for Toyota

A California-based defense lawyer for Toyota from 2003 to 2007. While working there, he was concerned that Toyota was withholding evidence from trial, and gathered documents to prove it.

Mr. Biller later provided the U.S. Congress with confidential documents that helped further investigations into if, when and how Toyota was breaking U.S. federal law (which we now know it was thanks in part to submission).

Toyota sued Mr. Biller for breach of confidentiality and won a $2.6 million settlement against him.[221]

Courtland Kelley – Once Head of GM Inspection Program

Courtland Kelly, once the head of a nationwide GM inspection program and a quality manager, helped to expose GM's lack of management response to major vehicle flaws.

As Head of GM's National Inspection Program, he "found flaws and reported them, over and over, and repeatedly found his colleagues' and supervisors' responses wanting. . . Frustrated with the limited scope of a recall of sport-utility vehicles in 2002, he sued GM under a Michigan whistleblower law. GM denied wrongdoing, and the case was dismissed on procedural grounds. *Kelley's career went into hibernation*; he was sent to work in another part of the company, and GM kept producing its cars."[222]

William McAleer – Tried to Warn GM Board Members

William McAleer, a former GM employee who conducted quality audits, was so concerned about the company's lack of support to fix known problems with their vehicles that he sued GM and finally wrote letters to each of GM's Board Members in 2002. In these letters, "he told the board it should stop shipments of unsafe cars, launch recalls, and revise quality controls to make the company "independent of corporate politics and cost-cutting concerns."[223]

He no longer works for GM.

Recalls Due to Real, Live Bugs

Curiously, in the course of finalizing the research for this book, I discovered that real, live bugs have actually been a problem for at least two car manufacturers, triggering recalls. In one case, a manufacturer actually *decided to update their software* to get rid of the problems caused by these little creatures in their engines – a very clever use of technology!

Mazda (April 2014) issued a voluntary recall of 42,000 2010 – 2012 Mazda 6 vehicles because spider webs were found to cause a fuel leak. The manufacturer had the same problem before, with 2009 – 2010 Mazda 6s, and had tried to fix the problem by installing a spring, but some spiders can still get past the device.

Sac spiders, it seems, like gasoline. These spiders crawl in and weave their webs in the evaporative canister vent line. This can cause a restriction in the line, put stress on the fuel tank and cause it to crack. When that happens, fuel can leak out.[224]

How is Mazda going to solve the problem of real, live bugs in the engine?

"Mazda is turning to a software update to solve the spider problem. New software — already installed on a number of vehicles — will make sure that the pressure won't reach dangerous levels, even if spiders manage to get inside the vents."

When asked why it's taken so long to resolve the problem, Jeremy Barnes, a spokesman replied, "Don't ask me, I'm afraid of the damn things." [225]

Toyota (2013) had a similar run-in with spiders, recalling 803,000 vehicles because, when spider webs clog drain holes on an air conditioning housing unit," water can leak onto the electronic module that controls the airbags, causing them to deploy unexpectedly.

In this case, Toyota didn't have a software fix for that. They managed to solve the problem the old-fashioned way: they "provided additional sealing of the air conditioning unit and added a protective layer."[226]

Acknowledgements

I'd first like to thank the people who had the courage to care and, at great personal sacrifice, came forward to tell the public about the goings on at Toyota and GM.

I'd also like to very much thank some key reporters who provided accurate information and took the time to ensure their readers had facts rather than simply regurgitate corporate press releases with a few bits and pieces of other information to fill out the article. Without your dedication, this book could not have been written:

* Junko Yoshida and Robert E. Charette at the *Electronic Engineering Times* wrote articles about Toyota that were well-researched, thorough and actually provided me with the first hint that something was really, really wrong in the world of automobile software and its governance. Junko Yoshida has, all along, paid particularly close attention to the Toyota twists and turns rather than just let a story pop up and disappear again.

* Jerry Hirsch of the *Los Angeles Times* tracked the tragic Mark Saylor crash in San Diego and provided solid

information to help us understand a tragedy that ultimately led to Toyota confessing that it did, indeed, have problems with its cars.

* Mike Colias of *Autoweek* always wrote excellent, informative pieces regarding recalls.

* Andy Greenberg of *Forbes Magazine* very early on highlighted in his work the serious risks of hacking into cars. And thank you, too, Charlie Miller and Chris Vasalek – two Helpful Hackers who proved it can be done.

* David McNeil of *The Japan Times* and Lazar Berman of *The Times of Israel* – thank you for bringing to my attention the incredible courage and compassion of Toyota Whistleblower Betsy Benjaminson. She is an inspiration and deserves more support and attention than she's getting in light of the Toyota lawsuit she is enduring.

Advocating for safer vehicles and road safety and calling attention to the flaws in today's regulatory framework, I'd like to thank:

* Clarence Ditlow of the The Center for Auto Safety and
* Sean Kane of the Safety Research and Strategies Inc.

Also, I'd like to thank a politician and some software experts for their very focused attention and expertise regarding the types of automobile technology issues facing us today:

* Thank you, U.S. Senator Ed Markey, for keeping a close watch on the auto industry and recognizing how very important it is that society protects itself from hacking, tracking and dangerous automobile software.

* Professor Phil Koopman, of Carnegie Mellon University, thank you for your very concise (for me, anyway!)

166

presentation on the <u>Toyota software findings</u> in the *Bookout v. Toyota* case.

- And Michael Barr of The Barr Group, where would we be without you and your team's expert analysis of the Toyota software code and your ongoing efforts to help organizations produce safe software? Thank you very, very much.

Additionally, without my "writing cheerleaders" and other people who have taken the time to review and comment thoughtfully on this book, I would not have been able to get it done. Being a bit of a technology geek, myself, you have helped reduce the geek-speak in this book and I'm sure the readers thank you for that, too. So, a big thank to:

- Cathy Bamford
- Sydney Clark
- Mark Jones
- Brian A. Kilgore
- Jana Schilder
- Urmas Sui
- Keltie Thomas
- And the other writers and lawyers who helped me pull this book together.

ABOUT THE AUTHOR

Patricia Herdman, author and founder of Glitchwatch.com and Test Matters, is an international consultant who runs enterprise-wide software test and business transformation programs for some of the world's largest and most technologically complex companies in both Europe and North America.

The author currently lives in Toronto, Canada.

Notes

[1] Berman, Lazar, "High Price but no regrets for Israel's gutsy Toyota whistle-blower," *The Times of Israel* online, March 13, 2014, http://www.timesofisrael.com/high-price-but-no-regrets-for-israels-gutsy-toyota-whistle-blower/, accessed February 23, 2015.

[2] "NHTSA Opens Investigation of 4.9 Million Chrysler Vehicles," Shepardson, David, *The Detroit News*, October 27, 2014. http://www.detroitnews.com/story/business/autos/chrysler/2014/10/27/nhtsa-opens-review-million-chrysler-vehicles/17993537/ accessed 2 February 2015.

[3] Hirsch, Jerry, "Toyota Admits Deceiving Consumers," *LA Times* online, March 19, 2014, http://articles.latimes.com/print/2014/mar/19/business/la-fi-toyota-settlement-20140320, accessed February 5, 2015.

[4] Mail Foreign Service, " 'There's no brakes . . . hold on and pray': Last words of man before he and his family died in a fiery Toyota Lexus crash," *The Daily Mail* online, February 3, 2010, http://www.dailymail.co.uk/news/article-1248177/Toyota-recall-Last-words-father-family-died-Lexus-crash.html, accessed December 30, 2014.

[5] Devine, Rory and Payton, Mari and Stickney, R., "CHP Officer, Family Killed in Crash," *NBC News* online, August 31, 2009, http://www.nbcsandiego.com/news/local/CHP-Officer-Family-Killed-in-Crash-56629472.html, accessed February 15, 2015.

[6] Hirsch, Jerry, op. cit.

[7] Yoshida, Junko, "Toyota Case / Vehicle Testing Confirms Fatal Flaws," *EE Times* online, October 31, 2014, http://www.eetimes.com/document.asp?doc_id=1319966, accessed December 30, 2014.

[8] Ross, Brian and Rhee, Joseph and Hill, Angela M. and Churchmach, Megan and Katersly, Aaron, "Toyota to Pay $1.2B for Hiding Deadly 'Unintended Acceleration'," *ABC News* online, March 19, 2014, http://abcnews.go.com/Blotter/toyota-pay-12b-hiding-deadly-unintended-acceleration/story?id=22972214, accessed March 11, 2015.

[9] Yoshida, Junko, "New 'Runaway Toyota' Case Tests DOJ's Integrity," *EE Times* online, September 12, 2014, http://www.eetimes.com/document.asp?doc_id=1323903, accessed February 15, 2015.

[10] Yoshida, Junko, "Toyota Case / Vehicle Testing Confirms Fatal Flaws," *EE Times* online, October 31, 2014, http://www.eetimes.com/document.asp?doc_id=1319966, accessed December 30, 2014.

[11] NASA "Technical Report to the National Highway Traffic Safety Administration (NHTSA) on the Reported Toyota Corporation (TMC) Unintended Acceleration (UA) Investigation, NESC Assessment #: TI-10-00618, January 18, 2011, p. 17.

[12] Ibid, p. 13.

[13] Valdes-Dapena https://nowtoronto.com/art-and-books/books/review-when-

the-doves-disappeared-by-sofi-oksanen/, Peter, "Pedals, drivers blamed for out of control Toyotas," CNN online, February 8, 2011, http://money.cnn.com/2011/02/08/autos/nhtsa_nasa_toyota_final_report/, accessed February 27, 2015.

[14] Safety Research & Strategies, Inc. – Toyota Timeline -
http://www.safetyresearch.net/toyota-sudden-acceleration-timeline

[15] Transcript, *Bookout v. Toyota*, Case No. CJ-2008-7969, State of Oklahoma, October 14, 2013, p. 111.

[16] Yoshida, Junko, "Timeline: Toyota Faces More Battles in Liability War," *EE Times* online, November 13, 2013, http://www.eetimes.com/document.asp?doc_id=1319985, accessed February 10, 2015.

[17] Barr, Michael, "An Update on Toyota and Unintended Acceleration," Embedded Gurus blog, October 26, 2013, http://embeddedgurus.com/barr-code/2013/10/an-update-on-toyota-and-unintended-acceleration/, accessed February 19, 2015.

[18] Tucker, Eric and Krishner, Tom, "Toyota to pay US $1.2-billion over safety problems, the largest ever US penalty for an automaker," *The Financial Post* online, March 19, 2014, http://business.financialpost.com/2014/03/19/toyota-to-pay-us1-2-billion-over-safety-problems-the-largest-ever-u-s-penalty-for-an-automaker/, accessed February 25, 2015.

[19] MacNeil, David, "Imperial Family's car woes sparked Toyota Whistleblower", *The Japan Times* online, http://www.japantimes.co.jp/news/2013/06/09/business/corporate-business/imperial-familys-car-woes-sparked-toyota-whistleblower/#.VOtP97DF92c, accessed February 23, 2015.

[20] Berman, Lazar, op. cit.

[21] Berman, Lazar, op. cit.

[22] Fletcher, Michael A., "Toyota reaches $1.2 billion settlement to end probe of accelerator problems," *The Washington Post* online, March 19, 2014, http://www.washingtonpost.com/business/economy/toyota-reaches-12-billion-settlement-to-end-criminal-probe/2014/03/19/5738a3c4-af69-11e3-9627-c65021d6d572_story.html, accessed December 28, 2014.

[23] "Toyota Posts Record $17.9 Billion Profit," *Business Insider* online, May 8, 2014, http://www.businessinsider.com/toyota-posts-record-179-billion-profit-2014-5, accessed February 19, 2015.

[24] Fletcher, Michael A., op. cit.

[25] Fletcher, Michael A., op. cit.

[26] Yoshida, Junko, "Honda Admits Software Problem, Recalls 175,000 Hybrids," *EE Times* online, July 10, 2014, http://www.eetimes.com/document.asp?doc_id=1323061, accessed February 10, 2015.

[27] Bar, Michael, "An Update on Toyota and Unintended Acceleration," Embedded Gurus blog, October 26, 2013, http://embeddedgurus.com/barr-code/2013/10/an-update-on-toyota-and-unintended-acceleration/, accessed February 19, 2015.

[28] Yoshida, Junko, "Toyota Case / Vehicle Testing Confirms Fatal Flaws," *EE*

Times online, October 31, 2014, http://www.eetimes.com/document.asp?doc_id=1319966, accessed December 30, 2014.

[29] "The Safety and Promise of Automotive Electronics: Insights from Unintended Acceleration", National Research Council Report to the Transportation Research Board, 2012, p. xii.

[30] Soble, Jonathan, "Honday, Grappling With Quality Problems, Will Replace Its President," *International New York Times* online, November 23, 2014, http://www.nytimes.com/2015/02/24/business/international/honda-replace-president-chief-executive-takanobu-ito.html?_r=0, accessed March 9, 2015.

[31] Cole, Robert E., "What Really Happened to Toyota?", *MIT Sloan Management Review* online, Summer 2011, p.7.

[32] Definition of Robot from Google online.

[33] Calo, Ryan, Assistant Professor at University of Washington School of Law, "Robotics and the Lessons of Cyberlaw," Legal Studies Research Paper No. 2014-08, dated "Forthcoming 2015", p.103. Can be downloaded here: http://papers.ssrn.com/sol3/papers.cfm?abstract_id=2402972, accessed February 19, 2015.

[34] Taylor, Peter Shawn, "Car hacking, the crime of tomorrow," *The National Post* online, February 7, 2012, http://news.nationalpost.com/2012/02/07/peter-shawn-taylor-car-hacking-the-crime-of-tomorrow/, accessed February 9, 2015.

[35] Motavalli, Jim, "Dozens of Computers That Make Modern Cars Go," *New York Times* online, February 4, 2010, http://www.nytimes.com/2010/02/05/technology/05electronics.html?_r=0, accessed March 10, 2015.

[36] Walters, W.L., "On-board Diagnostics for electronic emission control systems [at GM]," *Vehicular Technology Conference* in Michigan, September 1980, http://ieeexplore.ieee.org/xpl/articleDetails.jsp?arnumber=1622803, accessed March 2, 2015.

[37] C Programing software copied from: http://automon.donaloconnor.net, accessed 30 January 30, 2015.

[38] Cole, Robert E., op. cit., p.7.

[39] Yoshida, Junko, "Honda Admits Software Problem, Recalls 175,000 Hybrids," *EE Times* online, July 10, 2014, http://www.eetimes.com/document.asp?doc_id=1323061, accessed February 10, 2015.

[40] Bunkley, Nick, "Why you can't blame only GM for the Year of the Recall," February 13, 2015, *Automotive News* online, http://www.autonews.com/article/20150213/BLOG06/150219903/why-you-cant-blame-only-gm-for-the-year-of-the-recall?cciid=email-autonews-daily, accessed February 13, 2015.

[41] NHTSA complaint #10503698, filed March 23, 2013, http://www-odi.nhtsa.dot.gov/owners/SearchSafetyIssues, accessed February 13, 2015 with the search criteria "Mercedes" and "350".

[42] "BMW fixes security flaw in its in-car software," *Reuters* online, January 30, 2015, http://www.reuters.com/article/2015/01/30/bmw-cybersecurity-idUSL6N0V92VD20150130, accessed February 9, 2015.

[43] Pauli, Darren, "Can't afford a BMW or Roller? Just HACK its doors," *The Register (UK)* online, http://www.theregister.co.uk/2015/02/02/645k_too_much_for_a_wraith_then_just_hack_its_doors_open/, accessed February 9, 2015.

[44] Ramsey, Jonathan, "BMW recalling 7 series over inadvertently opening doors." *Autoblog* online, October 29, 2012, http://www.autoblog.com/2012/10/29/bmw-recalling-7-series-over-inadvertently-opening-doors/, accessed February 18, 2015.

[45] Atiyeh, Clifford, "Ford Recalls More than 1.3 Million Cars for Power Steering Fires and Floor Mats, *Car and Driver* online, May 30, 2014, http://blog.caranddriver.com/ford-recalls-more-than-1-3-million-cars-for-power-steering-fires-and-floor-mats/, accessed April 9, 2015.

[46] Ibid.

[47] Turkus, Brandon, "NHTSA investigating Ford's solution to May 2014 power steering recall", *Car and Driver* online, April 6, 2015, http://www.autoblog.com/2015/04/06/ford-escape-mercury-mariner-steering-recall-nhtsa-investigation/, accessed April 9, 2015.

[48] Phelan, Mark, "Software defects steals distinction from the 2015 Chevrolet Tahoe," *The Detroit Free Press* online, July 10, 2014, http://archive.freep.com/article/20140710/COL14/307100021/chevrolet-tahoe, accessed February 9, 2015.

[49] Ibid.

[50] Ibid.

[51] Colias, Mike, "GM recalls 303,000 new Chevy Silverado, GMC Sierra pickups for engine fire risk", *Automotive News* online, January 10, 2014, http://www.autonews.com/article/20140110/RETAIL05/140119956/gm-recalls-303000-new-chevy-silverado-gmc-sierra-pickups-for-engine, accessed February 5, 2015.

[52] Hlavaty, Craig, "Houston Man's Pickup Burns Half Hour After He gets his GM Recall Notice," *Houston Chronicle*, Janaury 17, 2014. http://www.chron.com/cars/article/Houston-man-s-pickup-burns-half-hour-after-he-5152922.php, accessed February 5, 2015.

[53] Shepardson, David, "GM Recalls 221,000 Cars for Breaking Problem," *The Detroit News*, September 20, 2014. http://www.detroitnews.com/story/business/autos/general-motors/2014/09/20/gm-recalls-cars-braking-problem/15950569/, accessed February 5, 2015

[54] "GM Recall List," *Autoblog*, October 22, 2014, http://www.autoblog.com/2014/10/22/general-motors-recall-list/, accessed February 15, 2015.

[55] Ibid.

[56] Ibid.

[57] Colias, Mike, "GM Halts 2015 Chevy Colorado, GMC Canyon deliveries to fix air bag flaw," *Automotive News* online, http://www.autonews.com/article/20141003/OEM11/141009891/gm-halts-2015-chevy-colorado-gmc-canyon-deliveries-to-fix-airbag-flaw, accessed February 9, 2015.

[58] Colias, Mike, "GM Recalls 2.7 million U.S vehicles for issues including

brake lights, brakes, wipers," *Automotive News* online, May 15, 2014, http://www.autonews.com/article/20140515/OEM11/140519931/gm-recalls-2.7-million-u.s.-vehicles-for-issues-including-brake, accessed February 15, 2015.

[59] NHTSA Recall Notice, April 3, 2014, http://www-odi.nhtsa.dot.gov/owners/SearchResults?searchType=ID&targetCategory=R&searchCriteria.nhtsa_ids=14V173000, accessed February 5, 2015.

[59] Charette, Robert, "GM Recalls 50,500 2011 Cadillac SRXs over Airbag-Related Software Glitch," *EE Times* online, June 22, 2011, http://spectrum.ieee.org/riskfactor/transportation/advanced-cars/gm-recalls-50500-2011-cadillac-srxs-over-airbagrelated-software-glitch, accessed February 9, 2015.

[60] Klayman, Ben and Von Ahn, Lisa, "U.S. safety regulators upgrade probe of Accords for air bag issue," *Reuters* online, August 4, 2014, http://www.reuters.com/article/2014/08/04/honda-usprobe-idUSL2N0QA0BN20140804, accessed March 12, 2015.

[61] Seetharaman, Deepa and Frank, Jackie, "Honda to recall 344,000 minivans in U.S. due to braking glitch," *Reuters* online, November 1, 2013, http://www.reuters.com/article/2013/11/02/autos-honda-recall-idUSL1N0IN00B20131102, accessed March 12, 2015.

[62] Yoshida, Junko, "Honda Admits Software Problem, Recalls 175,000 Hybrids", *EE Times* online, July 7, 2014, http://www.eetimes.com/document.asp?doc_id=1323061, accessed February 5, 2015

[63] Fahey, Mark, "Hyundai recalls 200,000 vehicles for power steering defect," CNN online, March 1, 2015, http://money.cnn.com/2015/03/01/autos/hyundai-recall/, accessed March 1, 2015.

[64] *Information Age, October 25, 2011,* http://www.information-age.com/technology/applications-and-development/1663983/jaguar-recalls-17500-cars-due-to-software-glitch# accessed on February 2, 2015.

[65] Wood, David A., "Mazda Tribute Recalled for Loss of Power Steering," CarComplaints.com, June 18, 2014, http://www.carcomplaints.com/news/2014/mazda-tribute-recalled-power-steering.shtml, accessed Feburary 5, 2015

[66] NHTSA Recall Notice, April 3, 2014, http://www-odi.nhtsa.dot.gov/owners/SearchResults?searchType=ID&targetCategory=R&searchCriteria.nhtsa_ids=14V173000, accessed February 5, 2015.

[67] Charette, Robert, op. cit.

[68] "Nissan recalls Infiniti hybrid sedans for software", Reuters, November 4, 2014, http://www.reuters.com/article/2014/11/04/us-nissan-infiniti-recall-idUSKBN0IO1DE20141104 accessed February 2, 2015

[69] Plungis, Jeff, "Faced with Recal, Tesla Asks Just What 'Recall' Means", *Bloomberg* online, January 17, 2014, http://www.bloomberg.com/news/articles/2014-01-17/elon-musk-s-english-lesson-speaks-to-quest-to-change-cars, accessed March 16, 2015.

[70] Hard, Gordon, "Faulty air bag electronics spark additional recalls," *Consumer Reports* online, February 3, 2015, http://www.consumerreports.org/cro/news/2015/02/faulty-air-bag-electronics-spark-additional-recalls/index.htm, accessed March 2, 2015.

[71] Spector, Mark, "Auto Makers Recalling 2.12 Million Cars and SUVs Over Air Bag Deployment", *The Wall Street Journal*, 31 January 2015, http://www.wsj.com/articles/auto-makers-recalling-2-12-million-cars-suvs-over-air-bag-deployment-1422723475?mod=WSJ_hp_LEFTWhatsNewsCollection, accessed February 2, 2015.

[72] Although Honda reported in 2010 there were no deaths associated with the recall of their vehicles due to the air bag problem, in 2014, the U.S. regulators fined Honda $70 million for failing to report over 1,700 deaths associated with their vehicles. It should be noted that in the U.S., car manufacturers are required to report deaths associated with their vehicles.

[73] Kanellos, Michael, "Software glitch stalls some Toyota hybrids," *CNET News* online, October 14, 2005, http://news.cnet.com/Software-glitch-stalls-some-Toyota-hybrids/2100-11389_3-5895574.html, accessed March 11, 2015.

[74] Conley, Margaret, "Toyota Officials Admit Problem with 2010 Hybrid Prius," *ABC News* online, February 4, 2010, http://abcnews.go.com/Blotter/RunawayToyotas/toyota-recall-toyota-admits-2010-hybrid-prius-problems/story?id=9744744, accessed February 12, 2015.

[75] Smith, Aaron, "Toyota Recalls 2.1 million vehicles," CNN online, February 12, 2014, http://money.cnn.com/2014/02/12/autos/toyota-prius-recall/, accessed February 17, 2015.

[76] Bennett, Jeff, "U.S. Fines General Motors for Missing Recall Deadline," *The Wall Street Journal* online, April 8, 2014, http://www.wsj.com/articles/SB10001424052702304819004579489370462753000, accessed March 13, 2015.

[77] Ewing, Steven J., "Chrysler recalls 350K vehicles over ignition switches," *Autoblog* online, September 2014, http://www.autoblog.com/2014/09/25/chrysler-dodge-jeep-350k-ignition-switch-recall/, accessed February 18, 2015.

[78] Krisher, Tom, "Chrysler recalls nearly 907,000 cars, SUVs to fix stalling and power mirror wiring problems," *The National Post* online, October 16, 2014, http://business.financialpost.com/2014/10/16/chrysler-recalls-nearly-907000-cars-suvs-to-fix-stalling-and-power-mirror-wiring-problems/#__federated=1, accessed February 12, 2015.

[79] Colias, Mike, "GM Recalls 2.7 million U.S vehicles for issues including brake lights, brakes, wipers," *Automotive News* online, May 15, 2014, http://www.autonews.com/article/20140515/OEM11/140519931/gm-recalls-2.7-million-u.s.-vehicles-for-issues-including-brake, accessed February 15, 2015.

[80] Gutierrez, Gabe and Cardella, Rich and Monahan, Kevin and Reynolds, Talisha, "Parents 'Boiling with Anger' After Daughter's Death in GM Car," NBC News online, March 14, 2014, http://www.nbcnews.com/storyline/gm-recall/parents-boiling-anger-after-daughters-death-gm-car-n52316, accessed February 11, 2015.

[81] Colias, Mike, op. cit.

[82] Shepardson, David, "Ford recalls 850,000 vehicles for electronic glitch," *The Detroit News* online, September 26, 2014, http://www.detroitnews.com/story/business/autos/ford/0001/01/01/ford-recalls-vehicles-electronic-glitch/16258347/, accessed February 15, 2015.

[83] Subaru Recalls, Cars101 online, http://www.cars101.com/recalls.html,

accessed February 15, 2015.

[84] Bowman, Zack, "VW recalls 2012-2013 Beetle models with leather seats over airbags," *Automotive News* online, November 20, 2012, http://www.autoblog.com/2012/11/20/vw-recalls-2012-2013-beetle-models-with-leather-seats-over-airba/, accessed February 11, 2015.

85 Google Definition, accessed January 30, 2015.

[86] Yoshida, Junko, "Toyota Case / Vehicle Testing Confirms Fatal Flaws," *EE Times* online, October 31, 2014, http://www.eetimes.com/document.asp?doc_id=1319966, accessed December 30, 2014.

[87] Yoshida, Junko, "Toyota Case / Vehicle Testing Confirms Fatal Flaws," *EE Times* online, October 31, 2014, http://www.eetimes.com/document.asp?doc_id=1319966, accessed December 30, 2014.

[88] Yoshida, Junko, "New 'Runaway Toyota' Case Tests DOJ's Integrity," *EE Times* online, September 12, 2014, http://www.eetimes.com/document.asp?doc_id=1323903, accessed February 15, 2015.

[89] Berman, Lazar, op. cit.

[90] Hirsch, Jerry, op. cit.

[91] "Regulators hired by Toyota Helped Halt Acceleration Probes," *Bloomberg* online, February 13, 2014, http://www.bloomberg.com/apps/news?pid=newsarchive&sid=atXvi2msqPOM, accessed March 11, 2015.

[92] Safety Research & Strategies, Inc. – Toyota Timeline - http://www.safetyresearch.net/toyota-sudden-acceleration-timeline

[93] Ibid.

[94] MacNeil, David, "Imperial Family's car woes sparked Toyota Whistleblower", *The Japan Times* online, http://www.japantimes.co.jp/news/2013/06/09/business/corporate-business/imperial-familys-car-woes-sparked-toyota-whistleblower/#.VOtP97DF92c, accessed February 23, 2015.

[95] Hirsch, Jerry, op. cit.

[96] Hirsch, Jerry, op. cit.

[97] Ross, Brian and Rhee, Joseph and Hill, Angela M. and Churchmach, Megan and Katersly, Aaron, "Toyota to Pay $1.2B for Hiding Deadly 'Unintended Acceleration'," *ABC News* online, March 19, 2014, http://abcnews.go.com/Blotter/toyota-pay-12b-hiding-deadly-unintended-acceleration/story?id=22972214, accessed March 11, 2015.

[98] "Toyota's 'Unintended Acceleration' has killed 89," *CBS News* online, May 25, 2010, http://www.cbsnews.com/news/toyota-unintended-acceleration-has-killed-89/, accessed February 15, 2015.

[99] Berman, Lazar, op. cit.

[100] NASA "Technical Report to the National Highway Traffic Safety Administration (NHTSA) on the Reported Toyota Corporation (TMC) Unintended Acceleration (UA) Investigation, NESC Assessment #: TI-10-00618, January 18, 2011, p. 17.

[101] Wallace, Ed, "Toyota: The Media Owes You an Apology", *Bloomberg* online, February 10,2011, http://www.bloomberg.com/bw/lifestyle/content/feb2011/bw20110210_848076.htm , accessed February 16, 2015.

[102] Berman, Lazar, op.cit.

[103] Toyota Settlement website, https://www.toyotaelsettlement.com/, accessed February 12, 2015.

[104] Risling, Greg, "Toyota Settlement the Largest in U.S. History Involving Automobile Defects," December 26, 2012, http://www.huffingtonpost.com/2012/12/26/toyota-settlement_n_2366720.html, accessed February 15, 2015.

[105] Risling, Greg, "Toyota Settlement the Largest in U.S. History Involving Automobile Defects," *The Huffington Post*, December 26, 2012, http://www.huffingtonpost.com/2012/12/26/toyota-settlement_n_2366720.html, accessed February 15, 2015.

[106] Yoshida, Junko, "Toyota Case / Vehicle Testing Confirms Fatal Flaws," *EE Times* online, October 31, 2014, http://www.eetimes.com/document.asp?doc_id=1319966, accessed December 30, 2014.

[107] Fletcher, Michael A., op. cit.

[108] Viswantha, Aruna and Ingram, David and Kayman, Ben, "Update 7: Toyota's $1.2 bn settlement may be model for U.S. probe into GM," *Reuters* online, http://www.reuters.com/article/2014/03/20/toyota-settlement-idUSL2N0MG0LS20140320, accessed February 23, 2015.

[109] Cole, Robert E., "What Really Happened to Toyota?", *MIT Sloan Management Review* online, Summer 2011, p.7.

[110] Berman, Lazar, op. cit.
EXCERPT FROM ABOVE ARTICLE: The engineers "sometimes admitted it was the electronic parts, the engine computer, the software, or interference by radio waves," Benjaminson wrote.

It took about a year of translating for Toyota until she really started "getting it," Benjaminson said. "I began to notice discrepancies between what the engineers were saying, which focused on electronics, and stuff from the executives, PR people, and lawyers, about how to fool the public."

"Efforts were made to find floor mats that would trap gas pedals and conveniently explain UA [unintended acceleration]. The R&D chief admitted that incompletely developed cars had gone into production and that quality control of parts was poor or nonexistent."

[111] Bensinger, Ken, "Toyota looks to settle sudden-acceleration lawsuits", Los Angeles Times online, December 12, 2013, http://articles.latimes.com/2013/dec/12/business/la-fi-toyota-settlement-20131213, accessed March 21, 2015.

[112] Ibid.

[113] Yoshida, Junko, "New 'Runaway Toyota' Case Tests DOJ's Integrity," *EE Times* online, September 12, 2014, http://www.eetimes.com/document.asp?doc_id=1323903, accessed February 15, 2015.

[114] English, Andrew, "Toyota's driverless car," *The Telegraph* online, October 25, 2015, http://www.telegraph.co.uk/motoring/car-manufacturers/toyota/10404575/Toyotas-driverless-car.html, accessed February 15, 2015.

[115] The figures regarding acceleration issues was determined by searching the U.S. Government NHTSA complaints database. http://www-odi.nhtsa.dot.gov/owners/SearchSafetyIssues, accessed February 12, 2015.

[116] Bill, Vlasic, "Toyota Agrees to Settle Lawsuits Tied to Accelerations," *New York Times* online, December 26, 2012, http://www.nytimes.com/2012/12/27/business/toyota-settles-lawsuit-over-accelerator-recalls-impact.html?_r=1&adxnnl=1&pagewanted=print&adxnnlx=1423580453-8XNu53I0F9a+YKTtJx42/A, accessed February 10, 2015.

[117] Lavrinc, Damon, "Google Poaches Deputy Directdor of National Highway Traffic Safety Administration," *Wired Magazine* online, Noveber 19, 2012, http://www.wired.com/2012/11/ron-medford-google-nhtsa/, accessed March 18, 2015.

[118] Bennett, Jeff, "GM Ignition Switch Death Claims Rise to 80," The Wall Street Journal online, April 6, 2015, http://www.wsj.com/articles/gm-ignition-switch-death-claims-rises-to-80-report-1428330029, accessed April 9, 2015.

[119] Higgins, Tim and Summers, Nick, "GM Recalls: How General Motors Silenced a Whistleblower," *Bloomberg Businessweek* online, June 18, 2014, http://www.bloomberg.com/bw/articles/2014-06-18/gm-recalls-whistle-blower-was-ignored-mary-barra-faces-congress, accessed Feburary 22, 2015.

[120] Bunkley, Nick and Colias, Mike, "GM fires engineers, executives after ignition switch recall probe," *Autoweek* online, June 4, 2014, http://autoweek.com/article/car-news/gm-fires-engineers-executives-after-ignition-switch-recall-probe, accessed Feburary 23, 2015.

[121] Penenberg, Adam, "GM's hit and run: How a lawyer, mechanic, and engineer blew open the worse auto scandal in history," *Pandodaily* online, October 18, 2014, http://pando.com/2014/10/18/gms-hit-and-run-how-a-lawyer-mechanic-and-engineer-blew-the-lid-off-the-worst-auto-scandal-in-history/, accessed February 22, 2015.

[122] Krishner, Tom (AP), "U.S. Judge sends key wrongful death case back to Georgia court," *CTV News* online, July 19, 2014, http://www.ctvnews.ca/business/u-s-judge-sends-key-gm-wrongful-death-case-back-to-georgia-court-1.1921625, accessed February 22, 2015.

[123] Bennett, Jeff, "U.S. Fines General Motors for Missing Recall Deadline," *The Wall Street Journal* online, April 8, 2014, http://www.wsj.com/articles/SB10001424052702304819004579489370462753000, accessed March 13, 2015.

[124] Gutierrez, Gabe and Cardella, Rich and Monahan, Kevin and Reynolds, Talisha, "Parents 'Boiling with Anger' After Daughter's Death in GM Car," *NBC News* online, March 14, 2014, http://www.nbcnews.com/storyline/gm-recall/parents-boiling-anger-after-daughters-death-gm-car-n52316, accessed February 11, 2015.

[125] Krishner, Tom (AP), "U.S. Judge sends key wrongful death case back to Georgia court," *CTV News* online, July 19, 2014, http://www.ctvnews.ca/business/u-s-judge-sends-key-gm-wrongful-death-case-back-to-georgia-court-1.1921625, accessed February 22, 2015.

[126] "GM loses bid to dismiss Melton wrongful death and fraud lawsuit in Georgia state court," Beasley Allen Law Firm, August 11, 2014, http://www.beasleyallen.com/news/gm-loses-bid-to-dismiss-melton-wrongful-death-and-fraud-lawsuit-in-georgia-state-court/, accessed February 21, 2015.

[127] Fisk, Margaret Cronin and Welch, David, "GM Switch Victim's Lawyer Says Ignition Files Show Cover-Up," *Bloomberg* online, March 16, 2015, http://www.bloomberg.com/news/articles/2015-03-16/gm-switch-victim-s-lawyer-says-ignition-documents-show-cover-up, accessed March 16, 2015.

[128] Bunkley, Nick and Colias, Mike, "GM fires engineers, executives after ignition switch recall probe," *Autoweek* online, June 4, 2014, http://autoweek.com/article/car-news/gm-fires-engineers-executives-after-ignition-switch-recall-probe, accessed February 23, 2015.

[129] "GM Settles with Family of Dead Ignition-Switch Victim," *Agence France Press*, March 16, 2015.

[130] Bennett, Jeff, op. cit.

[131] Elmer, Stephen, "GM Ignition Switch Death Toll Grows to 64," Autoguide online, March 9, 2015, http://www.autoguide.com/auto-news/2015/03/gm-ignition-switch-death-toll-grows-to-64.html, accessed March 9, 2015.

[132] Associated Press, "General Motors death toll rises to 57 in rashes linked to faulty ignition switch," *The Guardian* online, February 23, 2015, http://www.theguardian.com/us-news/2015/feb/23/general-motors-death-toll-ignition-switch, accessed February 28, 2015.

[133] Krishner, Tom (AP), op. cit.

[134] Lienert, Paul, "U.S. Safety Wathdog says 303 deaths linked to recalled GM cars," *Reuters* online, March 13, 2014, http://www.reuters.com/article/2014/03/14/us-gm-recall-casualties-idUSBREA2D02U20140314, accessed February 17, 2015.

[135] Sandler, Linda, "Bankruptcy Judge May Lift Shield for GM Over Ignition Switch Claims", *The Insurance Journal* online, March 27, 2015, http://www.insurancejournal.com/news/national/2015/03/27/362330.htm, accessed March 28, 2015.

[136] Barr, Michael, "A Look Back at the Audi 5000 Unintended Acceleration," Embedded Gurus blog, March 14, 2014, http://embeddedgurus.com/barr-code/2014/03/a-look-back-at-the-audi-5000-and-unintended-acceleration/, accessed February 25, 2015.

[137] McArdle, Megan, "NHTSA: No, Toyotas Do Not Accelerate Unless You Press the Accelerator", *The Atlantic* online, July 14, 2010, http://www.theatlantic.com/business/archive/2010/07/nhtsa-no-toyotas-do-not-suddenly-accelerate-unless-you-press-the-accelerator/59696/, accessed February 23, 2015.

[138] Ibid.

[139] White, Joseph B. and Searcey, Dionne, "The Audi Case Set Template for Toyota's Troubles", *The Wall Street Journal* online, March 12, 2010,

http://www.wsj.com/articles/SB10001424052748704349304575115952186305536, accessed February 24, 2015.

[140] Landler, Mark, "A Sputtering VW Aims Higher", *The New York Times* online, March 16, 2003, http://www.nytimes.com/2003/03/16/business/a-sputtering-vw-aims-higher.html, accessed February 23, 2015.

[141] Green, Gavin, "The Car Column: Audi is Making Room at the Top," *The New York Times* online, May 30, 1997, http://www.nytimes.com/1997/05/30/style/30iht-car.t_1.htmlm accessed February 24, 2015.

[142] Becker, Joshua L. and O'Donahue, Sarah [lawyers with Alston & Bird, LLP], "Pumping the Brakes on Unintended Acceleration Allegations," *Law360.com* online, February 2, 2015, http://www.law360.com/articles/616362/pumping-the-brakes-on-unintended-acceleration-allegations, accessed February 24, 2015.

[143] Greenberg, Andy, "Hackers Reveal Nasty New Car Attacks – With Me Behind the Wheel," *Forbes Magazine* online, July 24, 2013, http://www.forbes.com/sites/andygreenberg/2013/07/24/hackers-reveal-nasty-new-car-attacks-with-me-behind-the-wheel-video/, accessed February 5, 2015.

[144] Woollaston, Victoria, "Hackers hijack software in a CAR and remotely control the steering, brakes and horn using a laptop," *The UK Daily Mail*, July 13, 2013, http://www.dailymail.co.uk/sciencetech/article-2377841/Hackers-hijack-software-CAR-remotely-control-steering-brakes-horn-using-laptop.html, accessed March 15, 2015.

[145] Markey, Sen. Ed, *Tracking & Hacking: Security & Privacy Gaps Put American Drivers at Risk,*" February 2015, p.5.

[146] Taylor, Peter Shawn, "Car hacking, the crime of tomorrow," *The National Post* online, February 7, 2012, http://news.nationalpost.com/2012/02/07/peter-shawn-taylor-car-hacking-the-crime-of-tomorrow/, accessed February 10, 2015.

[147] Poulsen, Kevin, "Hacker Disables More Than 100 Cars Remotely," *Wired Magazine* online, March 17, 2010, http://www.wired.com/2010/03/hacker-bricks-cars/, accessed February 10, 2015.

[148] Holman, Pablos, *TEDxMIDWEST* (around 6 minutes in to the vi video). discussion), https://www.youtube.com/watch?v=hqKafl7Amd8, accessed 30 January, 2015.

[149] Markey, Sen. Ed, *Tracking & Hacking: Security & Privacy Gaps Put American Drivers at Risk,*" February 2015, p.5.

[150] Ibid., p.4.

[151] Ibid., p.2.

[152] Storm, Darlene, "Busted! Your car's black box is spying against you in court," *Computerworld* online, http://www.computerworld.com/article/2473294/data-privacy/busted--your-car-s-black-box-is-spying--may-be-used-against-you-in-court.html, accessed August 27, 2012.

[153] Greenberg, Andy, "Hackers Reveal Nasty New Car Attacks--With Me Behind The Wheel", *Forbes Magazine* online, July 24, 2013. http://www.forbes.com/sites/andygreenberg/2013/07/24/hackers-reveal-nasty-new-car-attacks-with-me-behind-the-wheel-video/, accessed February 5, 2015.

[154] I started writing the executive-focused book I mentioned in the preface at

the request of a concerned executive trying to do the right thing. He always got push-back from his Chief Information Officer whenever he tried to change how the organization managed (or didn't manage) quality. "Make it simple," he said, "and help them understand the consequences." Unfortunately, while researching for that book I stumbled upon the fact that we have serious issues to address in the automobile industry. But the fact remains: executives understand financial statements. That's their job. They need to also start understanding how the lack of quality impacts their financial statements in overt and less visible ways…as well as the lives of their customers. So that's my next book!

[155] Greenemeier, Larry, "Businesses Still Feeling Sting From Y2K Bug," *Information Week* online, January 10, 2000, http://www.informationweek.com/768/y2k.htm, accessed July 30, 2012. [link no longer live]

[156] Cole, Robert E., "What Really Happened to Toyota?", *MIT Sloan Management Review* online, Summer 2011, p.2.

[157] In the early 2000s, I led a successful quality improvement program for a bank that had the 6th most complex technology footprint in North America. In terms of technological complexity, NASA takes 1st place.

[158] Cole, Robert E., "What Really Happened to Toyota?", *MIT Sloan Management Review* online, Summer 2011, p.7.

[159] Ma, Jie and Hagiwana, Yukiko, "Honda Ditches Global Mid-term Sales Target After Recalls," *Bloomberg* online, February 13, 2015, http://www.bloomberg.com/news/articles/2015-02-13/honda-withdraws-global-midterm-sales-target-after-recalls, accessed February 14, 2015.

[160] "Honda Profits rise 7 percent on cost cuts, Japan sales", *Automotive News* online, July 29, 2014, http://www.autonews.com/article/20140729/OEM/140729858&template=printart, accessed February 15, 2015.

[161] Muller, Joann, "Toyota Admits Misleading Customers; Agrees to $1.2 Billion Criminal Fine," *Forbes Magazine* online, http://www.forbes.com/sites/joannmuller/2014/03/19/toyota-admits-misleading-customers-agrees-to-1-2-billion-criminal-fine/, accessed February 15, 2015.

[162] Berman, Lazar, op. cit.

[163] Yoshida, Junko, "Toyota Case / Vehicle Testing Confirms Fatal Flaws," *EE Times* online, October 31, 2014, http://www.eetimes.com/document.asp?doc_id=1319966, accessed December 30, 2014.

[164] Yoshida, Junko, "EE Times Community Weighs in on Toyota Case", *EE Times* online, November 7, 2013, http://to.outlet22.com/document.asp?doc_id=1320013, accessed March 18, 2015.

[165] Email to Author dated March 14, 2014 with link to the following video: http://www.mercedes-benz.ca/content/canada/mpc/mpc_canada_website/en/home_mpc/passengercars/home/owners/rescue_assist.html#_int_passengercars:home:core-navi:rescue_assist, accessed March 14, 2014.

[166] Transcript, *Bookout v. Toyota*, Case No. CJ-2008-7969, State of Oklahoma, October 14, 2013, p. 55, 56.

[167] Motavalli, Jim, "Sudden Acceleration: It's Not Just Toyota," *CBS News* Online, January 9, 2010, http://www.cbsnews.com/news/sudden-acceleration-its-bad-and-its-not-just-toyota/, accessed February 12, 2015.

[168] Google definition of fiduciary, accessed online on February 21, 2015.

[169] Lowry, Joan (AP), "Honda fined record $y70 million for not reporting death, injury complaints," *Global News* online, January 8, 2015, http://globalnews.ca/news/1763498/honda-fined-record-70-million-for-not-reporting-death-injury-complaints/, accessed March 3, 2015.

[170] Sullivan, Andy, "Obama asks Congress for dozens more auto-safety investigators," *Reuters* online, February 2, 2015, http://www.reuters.com/article/2015/02/02/us-usa-budget-autos-idUSKBN0L628W20150202, accessed February 25, 2015.

[171] Neuhauser, Alan, "Hyundai, Kia Face $300 Million Penalty," *US News* online, November 3, 2014, http://www.usnews.com/news/articles/2014/11/03/hyundai-kia-face-300-million-penalty-for-systematically-overstating-vehicle-mileage, accessed February 25, 2015.

[172] Lavrinc, Damon, "Google Poaches Deputy Directdor of National Highway Traffic Safety Administration," *Wired Magazine* online, Noveber 19, 2012, http://www.wired.com/2012/11/ron-medford-google-nhtsa/, accessed March 18, 2015.

[173] Calo, Ryan, "The Case for A Federal Robotics Commission," Center for Technology Innovation at Brookings, September 2014, p. 3.

[174] Mazmanian, Adam, "Do robots dream of big government?", FCW.com (a division of Public Sector Media Group in "The Business of Federal Technology"), September 15, 2014, http://fcw.com/Articles/2014/09/15/robotics-commission.aspx, accessed February 9, 2015.

[175] Leinert, Paul and White, Joe, "Google partners with auto suppliers on self-driving cars," *Reuters* online, January 14, 2015, http://uk.reuters.com/article/2015/01/14/autoshow-google-urmson-idUKL1N0UT21E20150114, accessed February 9, 2015.

[176] Lavrinc, Damon, op. cit.

[177] Miller, Claire Cain, "When Driverless Cars Break the Law", *New York Times* online, March 13, 2014, http://www.nytimes.com/2014/05/14/upshot/when-driverless-cars-break-the-law.html?_r=0&abt=0002&abg=1, accessed February 9, 2015.

[178] Summary of Toyota Conduct and Whistleblower Policy, 2008, http://www.toyotauk.com/media/2691-A%20Code_of_conduct_2008.indd.pdf, accessed February 28, 2015.

[179] Scope, Whistleblower Policy, Toyota in India, date unknown, p.1, http://www.toyotabharat.com/inen/about/whistle_blower_policy.aspx, accessed March 16, 2015.

[180] Bonstad, Amanda, "Toyota Nears Settlement With Blogging Translator," *The National Law Journal* online, http://www.nationallawjournal.com/id=1202716163775/Toyota-Nears-Settlement-with-Blogging-Translator, accessed February 23, 2015.

[181] Berman, Lazar, op. cit.

[182] "U.S. Senate panel OK's auto industry's whistle-blower incentive," *Autonews* online, February 26, 2015, http://www.autonews.com/article/20150226/OEM11/150229885/u-s-senate-panel-oks-auto-industry-whistleblower-incentive?cciid=email-autonews-daily, accessed February 26, 2015.

[183] Bennett, Jeff, "U.S. Fines General Motors for Missing Recall Deadline," *The Wall Street Journal* online, April 8, 2014, http://www.wsj.com/articles/SB10001424052702304819004579489370462753000, accessed March 13, 2015.

[184] Young, Angelo, "Ford Recalls: Another Batch of US Cars That Can Just Roll Away on Their Own," *International Business Times* online, August 15, 2014, http://www.ibtimes.com/hyundai-sonata-recalled-transmission-problems-risk-rolling-away-while-parked-1643446, accessed February 18, 2015.

[185] Automakers are not required by law to follow either the FDA Guidance for the Content of Premarket Submissions for Software Contained in Medical Devices nor the FAA National Policy on Software Assurance but both provide a sound framework when developing safety critical systems.

[186] Barr, Michael, "Unintended acceleration and other embedded software bugs", Embedded Gururs online blog, March 30, 2011, http://www.embedded.com/electronics-blogs/barr-code/4214602/Unintended-acceleration-and-other-embedded-software-bugs, accessed January 1, 2015.

[187] Online search of NHTSA complaints database with the words "suddenly accelerate" shows there are 750 Toyota-related complaints and 1,091 for all other manufacturers combined.

[188] Safety Research & Strategies, Inc. – Toyota Timeline - http://www.safetyresearch.net/toyota-sudden-acceleration-timeline

[189] Ibid.

[190] Report of The National Highway Traffic Safety Administration, "Technical Assessment of Toyota Electronic Throttle Control (ETC) Systems," February 2011, p. 2.

[191] Ibid., p. 3.

[192] "Regulators Hired by Toyota Helped Halt Acceleration Probes," *Bloomberg* online, February 13, 2010, http://www.bloomberg.com/apps/news?pid=21070001&sid=aTfVxj4_pJh4, accessed February 2, 2015.

[193] MacNeil, David, "Imperial Family's car woes sparked Toyota Whistleblower", *The Japan Times* online, http://www.japantimes.co.jp/news/2013/06/09/business/corporate-business/imperial-familys-car-woes-sparked-toyota-whistleblower/#.VOtP97DF92c, accessed February 23, 2015.

[194] "Regulators Hired by Toyota Helped Halt Acceleration Probes," *Bloomberg* online, February 13, 2010, http://www.bloomberg.com/apps/news?pid=21070001&sid=aTfVxj4_pJh4, accessed February 2, 2015.

[195] Hirsch, Jerry, op. cit.

[196] Furst, Randy, Toyota's Trial Begins Over Fatal 2006 Crash in St. Paul," Minneapolis *Star Tribune* online, January 8, 2015,

http://www.startribune.com/local/stpaul/287920781.html, accessed February 15, 2015.

[197] Ross, Brian and Rhee, Joseph and Hill, Angela M. and Churchmach, Megan and Katersly, Aaron, "Toyota to Pay $1.2B for Hiding Deadly 'Unintended Acceleration'," *ABC News* online, March 19, 2014, http://abcnews.go.com/Blotter/toyota-pay-12b-hiding-deadly-unintended-acceleration/story?id=22972214, accessed March 11, 2015.

[198] Hirsch, Jerry, op. cit.

[199] "Toyota's 'Unintended Acceleration' has killed 89," *CBS News* online, May 25, 2010, http://www.cbsnews.com/news/toyota-unintended-acceleration-has-killed-89/, accessed February 15, 2015.

[200] "Regulators hired by Toyota Helped Halt Acceleration Probes," Bloomberg online, February 13, 2014, http://www.bloomberg.com/apps/news?pid=newsarchive&sid=atXvi2msqPOM, accessed March 11, 2015.

[201] Berman, Lazar, op. cit.

[202] Risling, Greg, "Toyota Settlement the Largest in U.S. History Involving Automobile Defects," *β* online, December 26, 2012, http://www.huffingtonpost.com/2012/12/26/toyota-settlement_n_2366720.html, accessed February 15, 2015.

[203] Simanaitis, Dennis, "Toyota Admits Software Misread Crash Data; Bug Subsequently Fixed," *Road and Track* online, http://www.roadandtrack.com/auto-news/tech/toyota-admits-software-misread-crash-data-bug-subsequently-fixed/, accessed December 8, 2014.

[204] NASA "Technical Report to the National Highway Traffic Safety Administration (NHTSA) on the Reported Toyota Corporation (TMC) Unintended Acceleration (UA) Investigation, NESC Assessment #: TI-10-00618, January 18, 2011.

[205] Wallace, Ed, "Toyota: The Media Owes You an Apology", *Bloomberg* online, February 10,2011, http://www.bloomberg.com/bw/lifestyle/content/feb2011/bw20110210_848076.htm, accessed February 16, 2015.

[206] Cato, Jeremy, "Who Owes Toyota an Apology?", *The Globe and Mail* online, February 15, 2011, http://www.theglobeandmail.com/globe-drive/news/trans-canada-highway/who-owes-toyota-an-apology/article611752/, accessed February 16, 2015.

[207] Associated Press, "Toyota acceleration settlement, Judge Finalizing $1.6 Billion Deal," *ABC News* online, July 20, 2013, http://abc7.com/archive/9179727/, accessed February 16, 2015.

[208] Vlasic, Bill, "Toyota Agrees to Settle Lawsuit Tied to Accelerations," *The New York Times* online, December 26, 2012, http://www.nytimes.com/2012/12/27/business/toyota-settles-lawsuit-over-accelerator-recalls-impact.html?_r=0, accessed February 15, 2015.

[209] Yoshida, Junko, "Toyota Faces More Battles in Liability War," *EE Times* online, November 4, 2013, http://www.eetimes.com/document.asp?doc_id=1319985, accessed February 10, 2015.

[210] Yoshida, Junko, "Toyota Case / Vehicle Testing Confirms Fatal Flaws," *EE Times* online, October 31, 2014, http://www.eetimes.com/document.asp?doc_id=1319966, accessed December 30, 2014.

[211] Yoshida, Junko, "Toyota Faces More Battles in Liability War," *EE Times* online, November 4, 2013, http://www.eetimes.com/document.asp?doc_id=1319985, accessed February 10, 2015.

[212] Bill, Vlasic, op. cit.

[213] Risling, Greg, "Toyota Settlement the Largest in U.S. History Involving Automobile Defects," *The Huffington Post* online, December 26, 2012, http://www.huffingtonpost.com/2012/12/26/toyota-settlement_n_2366720.html, accessed February 15, 2015.

[214] *Chaudhary v. Toyota*, "Sudden Unintended Acceleration Resulting in Death," April 18, 2013, http://archive.news10.net/assetpool/documents/130422083327_Chaudhary%20v%20%20Toyota-Complaint-Filed%20(01464581)%20(2).pdf, accessed February 26, 2015.

[215] Yoshida, Junko, "Toyota Case: Single Bit Flip That Killed," *EE Times* online, October 25, 2013, http://www.eetimes.com/document.asp?doc_id=1319903, accessed February 10, 2015.

[216] Bensinger, Ken, "Toyota looks to settle sudden-acceleration lawsuits", Los Angeles Times online, December 12, 2013, http://articles.latimes.com/2013/dec/12/business/la-fi-toyota-settlement-20131213, accessed March 21, 2015.

[217] "Toyota's Recall Woes: Lexus Settlement," *The Economist* online, March 19, 2014, http://www.economist.com/blogs/schumpeter/2014/03/toyotas-recall-woes, accessed December 10, 2014.

[218] Hirsch, Jerry, op. cit.

[219] Viswantha, Aruna and Ingram, David and Kayman, Ben, "Update 7: Toyota's $1.2 bn settlement may be model for U.S. probe into GM," *Reuters* online, http://www.reuters.com/article/2014/03/20/toyota-settlement-idUSL2N0MG0LS20140320, accessed February 23, 2015.

[220] Yoshida, Junko, "New 'Runaway Toyota' Case Tests DOJ's Integrity," *EE Times* online, September 12, 2014, http://www.eetimes.com/document.asp?doc_id=1323903, accessed February 15, 2015.

[221] Bensinger, Ken, "Toyota wins $2.6 million judgement against former attorney," *Los Angeles Times* online, January 6, 2011, http://articles.latimes.com/2011/jan/06/business/la-fi-0106-toyota-biller-20110106, accessed February 23, 2015.

[222] Higgins, Tim and Summers, Nick, "GM Recalls: How General Motors Silenced a Whistleblower," *Bloomberg Businessweek* online, June 18, 2014, http://www.bloomberg.com/bw/articles/2014-06-18/gm-recalls-whistle-blower-was-ignored-mary-barra-faces-congress, accessed Feburary 22, 2015.

[223] Thompson, Marilyn and Klayman, Ben, "Exclusive: GM Board warned of serious problems by quality manager in 2002", *Reuters* online, June 20, 2014,

http://www.reuters.com/article/2014/06/21/us-gm-recall-idUSKBN0EW02O20140621, accessed February 22, 2015.

[224] Jensen, Christopher, "Corrosion Risk Prompts Ford Recall, and Spiders Clog Mazdas," *The New York Times* online, April 4, 2014. http://www.nytimes.com/2014/04/05/automobiles/corrosion-risk-prompts-ford-recall-mazda-stung-by-spiders.html?_r=0, accessed February 5, 2015.

[225] D'Orazio, Dante, "Is Your Car Infested With Spiders? Mazda Has a Software Update for That," *The Verge* online, April 6, 2014, http://www.theverge.com/2014/4/6/5587116/mazda-6-software-update-addresses-spider-problem, accessed February 5, 2015.

[226] Jenson, Christopher, "Toyota to Recall 803,000 Vehicles for Airbag Problem", October 18, 2013, *The New York Times* online, http://wheels.blogs.nytimes.com/2013/10/18/toyota-to-recall-803000-vehicles-for-air-bag-problem/, accessed February 5, 2015.

Made in the USA
Charleston, SC
13 October 2016